BRINGING THE *PMBOK® GUIDE* TO LIFE

BRINGING THE *PMBOK® GUIDE* TO LIFE

A Companion for the Practicing Project Manager

Frank P. Saladis, PMP
Harold Kerzner, Ph.D.

JOHN WILEY & SONS, INC.

This book is printed on acid-free paper. ♾

Published by John Wiley & Sons, Inc., Hoboken, New Jersey
Published simultaneously in Canada

Library of Congress Cataloging-in-Publication Data:
Saladis, Frank P.
Bringing the PMBOK guide to life: a companion for the practicing project manager/ Frank P. Saladis, Harold Kerzner.
p. cm.
Includes bibliographical references and index.
ISBN 978-0-470-19558-1 (cloth)
1. Project management. I. Kerzner, Harold. II. Title.
HD69.P75.S25 2009
658.4'04—dc22

2008052044

Printed in the United States of America

10 9 8 7 6 5 4 3 2 1

Contents

Foreword

Every company has projects that are uniquely characterized by the size and nature of the business, the length of the project, whether for internal or external clients, whether or not a profit margin is included, and the project's strategic importance. Creating a project management methodology to encompass all of these characteristics is difficult, and even more complex is the attempt to create an enterprise project management methodology capable of use on all projects.

Most companies do not have the resources to research/benchmark other companies on what works and what fails, and this holds true even if the company possesses a project management office (PMO). As such, companies must rely on established project management standards.

The *PMBOK® Guide* provides the necessary framework and standards for project management. The real value in the use of the *PMBOK® Guide* lies in the guidance it provides companies in how to manage projects, irrespective of the characteristics. The *PMBOK® Guide* is also invaluable as the primary source for creating project management methodologies.

There are three ways to manage projects; the right way, the wrong way, and the *PMBOK® Guide* way. Not all of the information contained in the *PMBOK® Guide* will be applicable to all companies, nor should a company force all of the material to apply. The value in the *PMBOK® Guide* is that—it is just a "guide."

Every three or four years, hundreds of PMI members pool their intellectual knowledge to update the *PMBOK® Guide* based on current best practices. Companies should rely on the *PMBOK® Guide* for both current and future applications of project management best practices.

Harold Kerzner, Ph.D.
Executive Director for Project Management
The International Institute for Learning

Acknowledgments

My sincere thanks to my many mentors and friends in project management for their continued support—especially Dr. Harold Kerzner for his continued encouragement and my associates at the New York City Chapter of the Project Management Institute, my colleagues in the PMI Leadership Institute, the staff members at the International Institute for Learning, and my family for their support and understanding while dealing with the challenging life of a project manager.

Frank P. Saladis, PMP

Introduction

The field of project management is actually a collection of skills, tools, knowledge, techniques, lessons learned, insights, and observations gained from just about every industry and profession. Project managers develop skills in communications, financial management, conflict resolution, negotiation, planning, organizing, leading, and much more. There is a huge base of knowledge available for project managers. That base of knowledge is referred to as the *body of knowledge of project management*. This is such a large base of information that it would be impossible to include every element of project management in one book. Project managers are involved in strategic planning, disaster recovery, research, green technology, information technology, construction, and pretty much anything that involves people and plans. Most project managers are in a constant search for best practices and lessons learned that will help them improve how projects are implemented and to increase the probability of success.

The continued search for information has generated book after book about the subject. Each author is looking to introduce a new technique or find some new way to manage projects more effectively. Some new ideas are introduced, but much of the material that is available is a reissue of what is already known, just packaged slightly differently. The importance here is that project managers continue to develop the profession through new books, articles, and presentations, and they are using their creativity, innovation, and passion for the profession to add to the existing and ever-growing body of knowledge.

This vast collection of information and knowledge handed down year after year, with a continuous stream of new data coming in, requires some type of organized system or standard. The Project Management Institute, utilizing project managers from a wide range of industry including government, the private sector, and nonprofit organizations, have developed a standard from which project managers can create project management methodologies and benefit from the knowledge of others in the profession. That standard is known as *A Guide to the Project Management Body of Knowledge*. The standard, ANSI/PMI 99-001-2004, is currently in its 4th Edition. The next version was issued on December 2008.

The *PMBOK® Guide*, as it is commonly known, was developed and will continue to be developed by volunteers who offer their time and expertise in the continuing pursuit of higher levels of quality and improvement. The *PMBOK® Guide* can be found in the libraries of thousands of project managers worldwide, and has been translated into at least eight languages. It is used as part of the study and preparation for the PMP exam (Project Management Professional) offered by the Project Management Institute (PMI) and as reference for planning projects or to develop customized project management methodologies for hundreds of organizations and project management offices (PMOs).

The *PMBOK® Guide* is extremely useful to project managers at any level of an organization and for any type of project—from short-term, limited deliverable type projects to large, complex undertakings. It is important to note that it is a guide and is not all-inclusive. The *PMBOK® Guide* provides a solid basis for planning, and it introduces key processes and provides a framework for understanding project management. There are many views and perspectives about the *PMBOK® Guide,* and there is probably a fair amount of disagreement about how it should be used, and even some of its contents. Regardless of viewpoint, it is a valuable addition to any project manager's library.

THE *PMBOK® GUIDE*—THE BASICS

The *PMBOK® Guide* provides a foundation from which project plans and project management office (PMO) methodologies or enterprise wide processes can be developed. The first step is to become ***PMBOK® oriented.*** This means to become familiar with how the information in the document is presented and what the specific terminology used in the *PMBOK® Guide* means. This is particularly important because, although the terminology is used commonly across many industries, there are differences in meanings that could cause some confusion and miscommunication. An important item to remember is that the *PMBOK® Guide* is just that, a guide. It is not the entire project management body of knowledge condensed into about four hundred pages. The *PMBOK® Guide* is a representation of many best practices in project management that have evolved over the years and *may,* let me emphasize *may,* be used (meaning that there are many options, depending on the type of project) to manage a project successfully.

Chapters One through Three of the *PMBOK® Guide* provide an introduction to the reader and establish the basic framework from which the remaining chapters have been developed. These chapters introduce the forty-two project management processes that are mapped to the nine knowledge areas and five process groups. The remaining chapters describe the nine knowledge areas of project management.

A BRIEF DESCRIPTION OF THE PROJECT MANAGEMENT KNOWLEDGE AREAS

Integration management. This knowledge area emphasizes the generally accepted role of a project manager—coordination and bringing all the pieces (the deliverables of the project) together.

Scope management. Defining and determining what work must be done. It includes setting clearly defined project objectives, defining major project deliverables, and controlling changes to those deliverables. Scope management includes creating the work breakdown structure (a breakdown of the major project elements to improve planning and to assist in overall project control) to understand the complexity of the project.

Time management. The project is further defined through activity definition, sequencing of activities, estimating the duration of activities, determining the critical path, schedule development and managing schedule and time issues effectively.

Cost management. This involves estimating all project costs, budgeting costs over time, and controlling costs throughout the project life cycle.

Quality management. This area includes developing plans to ensure that requirements are met, establishing a quality policy, understanding quality principles introduced by quality experts, developing quality assurance processes, and controlling the quality of all project deliverables.

Human resource management. This involves identifying project stakeholders, developing the project team, motivating the team, understanding management styles, and organizational structure.

Communications management. This involves planning for and distributing information correctly and to the appropriate stakeholders, performance reporting, managing stakeholders, and developing processes to ensure effective transfer of information. Communications management includes developing an understanding of the communications sender–receiver model (transmitting messages, possible message distortion, and feedback loops).

Risk management. This area includes identifying potential project risk events, using qualitative analysis (expert judgment and experience) or quantitative analysis (using mathematical models and computer simulations) to prioritize potential risks, respond to risk situations, and develop risk monitoring and controlling processes.

Procurement management. This includes determining what goods and services should be purchased or developed internally by an organization, planning purchases and, developing procurement documentation such as requests for proposals (RFP). It also involves determining appropriate contract types, negotiating terms, selecting sellers, managing contracts through implementation, and then managing project closure and contractual closure.

The purpose of this book is to create a bridge between the larger, broader project management body of knowledge, the *PMBOK® Guide*, and the practicing project manager. This book is not all-inclusive regarding the subject of project management, but will provide some additional knowledge, clarification of terms, suggested approaches for the use of project management tools and techniques, and templates developed directly from information provided in the *PMBOK® Guide*. The book also provides suggestions and study tips that will assist in preparing for the PMP exam and includes a Project Plan Accelerator (PPA) that can be used with the *PMBOK® Guide* by project managers and project teams to develop plans that are specifically tailored to meet the needs of the client or sponsoring organization.

Chapter One

Why Do Project Managers Need This Book?

The main purpose of this book is to create a connection between what is referred to as "the body of knowledge of project management," the Project Management Institute's *A Guide to the Project Management Body of Knowledge®* (better known as the *PMBOK® Guide*), and the needs of the practicing project manager. The book is not specific to any industry. It is intended to provide additional information regarding the processes associated with managing projects and offer the reader other perspectives about the discipline including a practical and useful explanation of many of the tools, techniques, and processes described in the *PMBOK® Guide*. There are countless other books about project management that include suggested methodologies, templates, defined processes and procedures, and best practices—and many are excellent sources of reference. The focus of this book is on the practitioner, especially those who manage projects of moderate complexity.

This book is intended to make a direct connection between many of the terms and specific tools, techniques, inputs, and outputs described in the *PMBOK® Guide* and the common needs of today's practicing project managers. The objective is to provide a quick reference and a source of information that translates tools and techniques into useful templates, actionable steps, clarified processes, and common-sense approaches to managing a project.

The material and references in this are designed to be of interest to the newly appointed project manager as well as the seasoned professional. Experienced project managers may find the material helpful in further developing well-defined practices they are currently using, or the material may spark a new level of creativity and innovation that will take project management to a new level of efficiency. This book will help project managers and students of project management differentiate between generally practiced processes and those processes that are specific to a particular organization.

I have heard many people refer to the *PMBOK® Guide* as the "PMI Way" or refer to project management methodology in terms of "the Right Way, the wrong way, and the "PMBOK® way." The *PMBOK® Guide* does not suggest a right or wrong way and, in my opinion, there is no specific "PMI Way" or "PMBOK® Way." Everything in the *PMBOK® Guide* originates from the knowledge of hundreds of project managers who have decided to share their experiences across a very broad spectrum of projects and industries. This shared knowledge has been organized into the nine knowledge areas and 5 processes which provide the basis for developing organizational project management methodologies. Simply stated, there are projects that are managed well and those that are not managed well regardless of preferred methodology. If the methodology was developed using logic, common sense and the basic principles of project management along with a connection to the organization's culture, work ethic, business goals and capabilities then that would be the "right" methodology for the organization. Chances are that many of the project management steps and processes found in any organizational methodology can also be found in the *PMBOK® Guide* or can be directly related to *PMBOK® Guide* processes. I believe that most project managers will agree that in the discipline of, or if you prefer, the profession of **project management** the general principles are the same, regardless of what type of project is being managed. These general principles are then modified and combined with different approaches specific to the organization that can be used to achieve the common goals of on time completion, remaining within budget, and according to specifications. (there are other goals to consider but these are the goals most people associate with project success). The *PMBOK® Guide* provides a solid *framework for developing a methodology that would have a high probability of being accepted by the stakeholders of an organization.* It may not provide the extreme levels of detail that may be needed to develop a complete methodology, but it does provide the project manager with a reference point developed through the experience of hundreds of project managers over many years.

The purpose of this book is to bridge the needs of the project manager with the vast stores of knowledge about project management and to encourage project managers to expand their knowledge about the profession, challenge some of the processes in place and develop newer, more efficient ways of managing projects in a world filled with complex projects and new opportunities. This book will, I hope, encourage more thought about how to manage projects more effectively and open up the creative minds of project managers who can widen the project management information highway.

THE *PMBOK® GUIDE*, THE BODY OF KNOWLEDGE OF PROJECT MANAGEMENT, AND THE CONNECTION TO MANAGING PROJECTS

The *PMBOK® Guide* is a collection of tools, techniques, and processes developed by project managers engaged in projects across every type of business and industry. Project managers have offered their time and their expertise voluntarily to produce a consensus-based standard for managing a project. It is a valuable source of information for project managers, project management professionals, and future project managers, regardless of discipline. The *PMBOK® Guide* provides the basis for developing project management methodologies and can be found in the library of literally thousands of project management offices and practicing project managers. The challenge is to take the information provided in the *PMBOK® Guide* and apply it most effectively and appropriately to an actual project environment. This book focuses on specific techniques, explanations of terms, and application of tools that will enable project managers to effectively adapt the principles and processes described in the *PMBOK® Guide* to the practical world of project management. These techniques transform the *PMBOK® Guide* from a framework and standards reference to a sharpened tool in the project manager's toolbox.

Project management and the processes included within it definition is actually a collection of knowledge from every business discipline, and that knowledge base is in a continuous state of growth. This knowledge encompasses what has been learned through years of managerial experience, studying human behavior, analyzing relationships between organizations, engaging in strategic planning, managing conflict, performing financial planning, and understanding organizational structure and overall organizational performance. The greater, more complex, project management body of knowledge is basically a repository of information and best practices gathered from and covering every aspect of business and organizational management. The *PMBOK® Guide* organizes that knowledge in a logical format that is not intended to be viewed as a set of rules and regulations. It provides a basis for developing customized methods that will assist in meeting an organization's project and strategic objectives through an effective project management process. The lessons learned from completed projects are documented and shared (whenever possible) through networking with other practitioners, articles, books, and other forms of knowledge transfer and are added to the larger body of knowledge in a continuous and ever expanding cycle.

THE FIRST STEP: UNDERSTANDING HOW THE *PMBOK® GUIDE* WORKS

Many project managers consider the *PMBOK® Guide* to be the method for managing projects and carry it around as if it were some type of project management law book that must be followed. They profess that failure to follow the *PMBOK® Guide* will result in certain project failure and costly customer dissatisfaction. This type of thinking may result in a very inflexible approach to the management of the project and an attempt to force a technique or a process that is inappropriate for the project or that will cause unneeded work and possibly team frustration. This approach will, in many cases, result in resistance or even rejection of the *PMBOK® Guide* by management and the project team.

The first step in bringing the *PMBOK® Guide* to life is to understand that it is a guide (see Figure 1.1). The processes, tools, and techniques described in the document are meant to be considered and applied when *appropriate*. An inflexible attitude and approach in the use of the information provided in the *PMBOK® Guide* may result in considerable resistance by

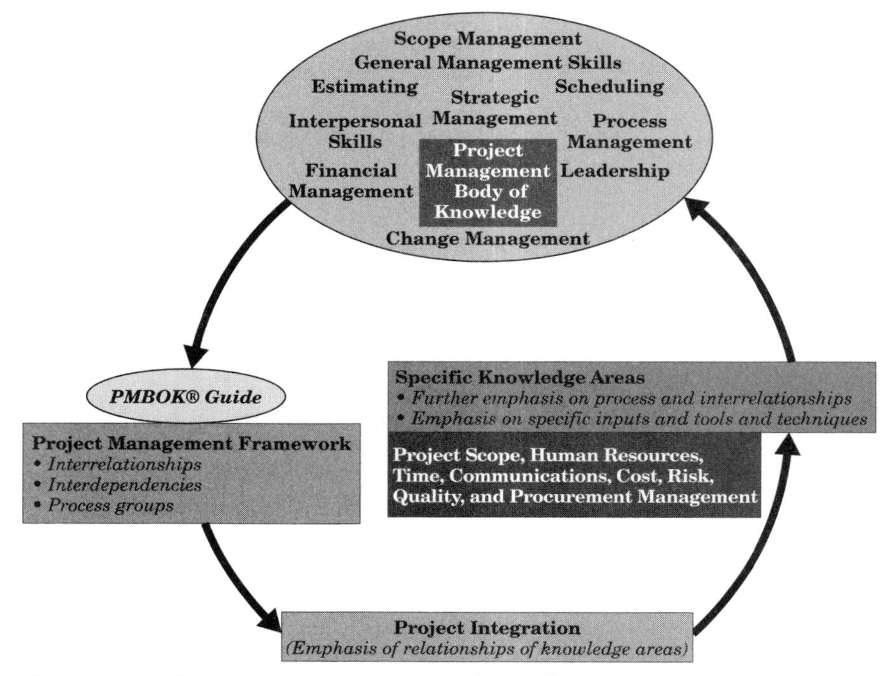

Figure 1.1 Project Management Life Cycle.

the project team or other stakeholders, a perception of inflexibility among team members about the project manager, challenges about the process, and possibly some undesired assistance from upper management or the project sponsor. To prevent this, it is important for the project manager and team to develop an understanding of how the *PMBOK® Guide* has been developed and structured and to review *each input, tool and technique,* and *output* within each process group described in each knowledge area and to understand their relationships and interdependencies during project planning and implementation:

Inputs. For all practical purposes, inputs are "things." They are generally deliverables (tangible work outputs) and are, in many cases, the outputs of other business or

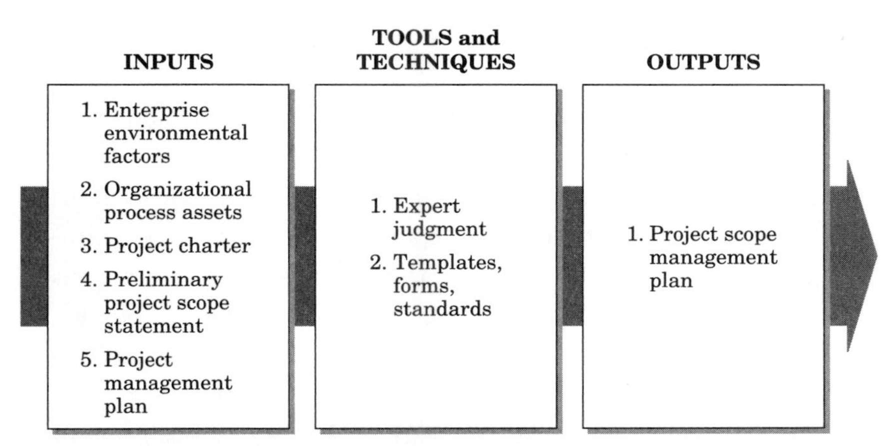

Figure 1.2 Process Flow for Scope Planning. *PMBOK® Guide*—Third Edition.

project management processes. Let's consider these to be *nouns*. They describe something tangible. These deliverables have specific names and were developed through the efforts of one or more people. It's important to understand what these inputs are, why they are required, and where they originate. Referring to Figure 1.2, enterprise environmental factors and organizational process assets are inputs to scope planning. Enterprise environmental factors include but are not limited to organizational culture, government standards, and infrastructure. Organizational Process assets include such items as policies, communication requirements, templates, project closure guidelines, and change control procedures. The project team should become aware of all environmental factors and organizational process assets that may affect the planning and control of the project. As an example, awareness of the organization's approach to risk—risk averse or aggressive risk taker (an environmental factor)—will impact how a plan is developed and how decisions about critical project issues will be made. This example emphasizes the importance of having a thorough knowledge of all inputs to each process group.

Tools and techniques. These are the specific actions and supporting items that allow us to utilize the identified inputs required to meet project needs. To further explain the tools and techniques component of the process a simple analogy would be that these tools and techniques are what you would find in a project manager's toolbox. They are a set of enablers and are used to shape and from the inputs into useful outputs when handled correctly. Consider tools and techniques to be *input processors*, very much like a blender or food processor in a kitchen. Once you have determined what your objective is—a cake, a tossed salad, or a special sauce—the appropriate inputs or "ingredients" are gathered. The reason why the inputs are needed should be clearly understood—you should know why they are being used. The inputs are processed using the selected tools or techniques to produce a desired result (the outputs). This is a relatively simple explanation, but the objective of a project methodology should be to keep it as simple as possible, use the appropriate tools (making sure you know how to use them) and avoid processes that will overwhelm the project team or cause confusion.

Outputs. The results produced through the use of the tools and techniques become outputs. Outputs become tangible items or deliverables that will be used as inputs to another process or will be finalized for handoff and use by the intended customer or stakeholder.

The *PMBOK® Guide*, through these processes, emphasizes that in the project environment, the customer is actually the next person in the process and not just the organization or entity that will receive and pay for the product of the project. The inputs, tools

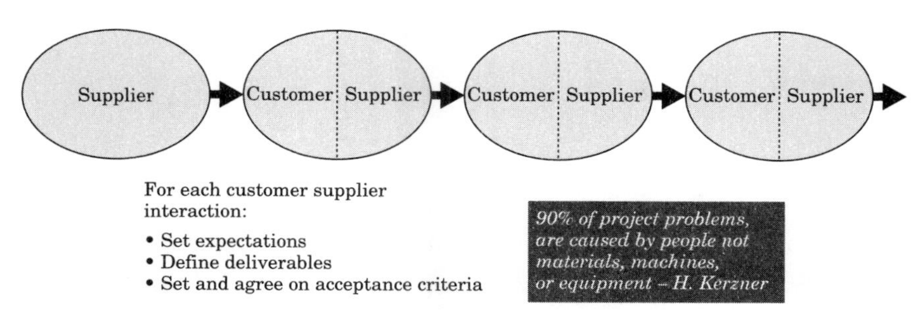

Figure 1.3 Project Customer–Supplier Model.

and techniques, and outputs create a project **customer–supplier model**. This model (Figure 1.3) refers to the fact that outputs of a process, the deliverables that are created using the tools and techniques, may be handed off to another project team member for further processing. This makes that particular team member or functional manager a customer. This handoff starts the next cycle of inputs, tools and techniques, and outputs. Maintaining focus on this model throughout the project life cycle will improve overall quality and planning efficiency, and should reduce project rework, resulting in improved overall performance and greater probability of success. It is important to note that each knowledge area in the *PMBOK® Guide* includes an overview of the inputs, tools and techniques, and outputs associated with that specific knowledge area. These overview charts present a type of "roadmap" of processes and illustrate relationships between process groups and other knowledge areas. The inputs, tools and techniques, and outputs found in one knowledge area may be included in the processes associated with other knowledge areas as part of a different process and different targeted result. This is a further indication of the integration of processes, and that concurrent planning of each project element or knowledge area is the norm for most projects. These overview charts also provide a means to demonstrate clearly how each knowledge area is connected and *integrated* within the entire project management planning process and create a true systems approach to planning and executing the project.

KNOWLEDGE AREAS AND PROCESS GROUPS

Project managers are aware that most organizations achieve their objectives through operations, programs, and the completion of projects. Strategic goals are established, operations procedures are formed, and the resources of the organization are assigned to specific responsibilities—either to complete projects or to keep the organization operating. Programs are developed for managing long-term services, applications, and other essential elements for the business. Projects are approved and chartered to support programs, create new or update products or services and at some point in time to replace a program that has outlived its usefulness.

Projects and programs are a part of everyday life. Although many people don't realize it, project management in some form is used by almost everyone, regardless of profession or occupation. Projects include planning a vacation or remodeling a home, developing a new vehicle, constructing a new building, or sending a team of astronauts to the international space station. Governments utilize formal project management to update infrastructure, the military uses project management to plan battle campaigns, private businesses use project management to improve their competitive edge. Regardless of type of project or industry, the basic principles of project management are the same: Determine what must be accomplished or what problem must be solved, select the best, most cost-effective approach, pick a team, obtain funding, plan the project, execute the plan, manage variances, and eventually close out the project.

It sounds simple when stated like that, but any experienced project manager will tell you that it is much more involved. The challenge is to manage projects, especially large, complex undertakings, with smoothness, control, and without the bureaucracy and complications that seem to be included in the perceptions of some people when project management is discussed.

The ideas and principles of project management date back thousands of years. In the past fifty to sixty years, project management has become modernized and more formalized, and new tools have been added to manage projects as they became more complex. Today, there are dozens of planning software tools to choose from, training programs to hone skills, and systems that can be used at the enterprise level to more effectively integrate project activities with the operations of the organization. Project management processes are now becoming more and more common at the higher management levels of organizations and are considered by many to be a key component of strategic planning.

Project Management Processes

Strategic planning is used to determine the direction an organization should be moving in, create scenarios about the future of a business, and forecast out the condition or health of an organization one to three years. In some cases, it may be a longer view. The strategic plan is used to explain how the organization is going to achieve its desired goals and how it will know if it actually reached them. The focus of a strategic plan is usually on the entire organization, while the focus of a project plan is usually on a particular product, service, or program that will support the higher-level strategic plan. It is important to have a process in place that will provide the organization with the means to achieve the goals of the strategic plan. That means of achievement is project management. An organized approach with a specific and defined process that is communicated to the stakeholders within the organization, supported by management, and followed through with minimal deviation should result in success at the project level and the organizational level.

Trying to start from the ground level and attempting to develop internal processes that can be used to achieve objectives is a long and challenging process. Many organizations start with an **ad hoc approach,** meaning "for the specific purpose, case, or situation at hand and not for any other reason." For unexpected situations and items that require a quick turnaround, the ad hoc approach may be appropriate and is probably used frequently within many companies. But the ad hoc approach does not address the longer-term needs of an organization and certainly does not promote the development of more stabilized and consistent processes. As companies mature in the management of projects, they begin to seek best practices that will help them reduce the learning curve and produce results at a faster pace. The *PMBOK® Guide* provides a basis for improved project planning by explaining key principles of project management and organizing project planning into a logical system that is integrated by processes. The nine knowledge areas of the *PMBOK® Guide*—integration, scope management, time management, cost management, quality management, Human Resources Management, communications management, risk management, and procurement management—are integrated through five processes—**initiating, planning, executing, monitoring and controlling,** and **closing.** The basic process groups are generally included in every phase of a project and provide the following:

1. *Initiating*. The purpose of the project is defined, the project is authorized, project managers are selected and teams are formed. The subprocesses within this group

are utilized in all project phases and serve as a way to determine if a project should proceed to the next phase.

2. *Planning.* The project objectives are established and developed through a progressive process of definition sometimes referred to as *progressive elaboration.* The project scope is defined in more detail, schedules are developed, procurement plans are defined, risks are identified, resource needs are identified, quality measures are determined, and communications plans are agreed upon.
3. *Executing.* The work defined in the project plan is initiated. The project manager coordinates activities, manages resources and subcontractors, procures goods and services, and observes project and team performance.
4. *Monitoring and Controlling.* The project team and project manager observe the work, analyze results, identify variances, and determine solutions and implement corrective action when necessary.
5. *Closing.* As each phase of a project is completed, a closing process is utilized to ensure that all deliverables have been completed to the satisfaction of the client and other stakeholders. At project completion, an analysis of project performance is conducted to ensure that all contractual obligations have been met, all work orders are completed, and any lessons learned are documented for use on future projects. The closing processes ensure formal acceptance of the project deliverables and that all project activities are finalized. The closing of a successful project should also include some type of recognition event for the project team. Recognition in some way is essential to maintain team loyalty. Project managers in many organizations will work with the same team members on multiple projects, and a good relationship with the team members will help to

minimize conflict and assist in influencing and motivating team members throughout the project life cycle.

These processes generally overlap in each phase of a project and interact throughout a project or phase, bridging and integrating elements of the nine knowledge areas. Process interactions are described in terms of the following:

- *Inputs* are documents, plans, designs, deliverables—generally something tangible.
- *Tools and techniques* are actions and approaches that are applied to the inputs.
- *Outputs* are documents, products, deliverables—tangible results of the applied actions. The outputs of many of these process interactions become inputs to other processes.

The *PMBOK® Guide* discusses inputs, tools and techniques, and outputs in each knowledge area chapter. As you proceed from chapter to chapter you will see many of the same inputs and tools and techniques being used in different processes. This is a clear indication of the integrated nature of projects. The inputs and tools and techniques described in the *PMBOK® Guide* may not be used in every project, but developing an understanding of why they may be needed will assist a project manager in minimizing the omission of critical project elements and planning components.

The Purpose of Inputs

Project managers should have an understanding about why an input is needed. Each input is basically an ingredient that is needed to complete a process and bring about a result. It is important to have some general knowledge of the origin of the

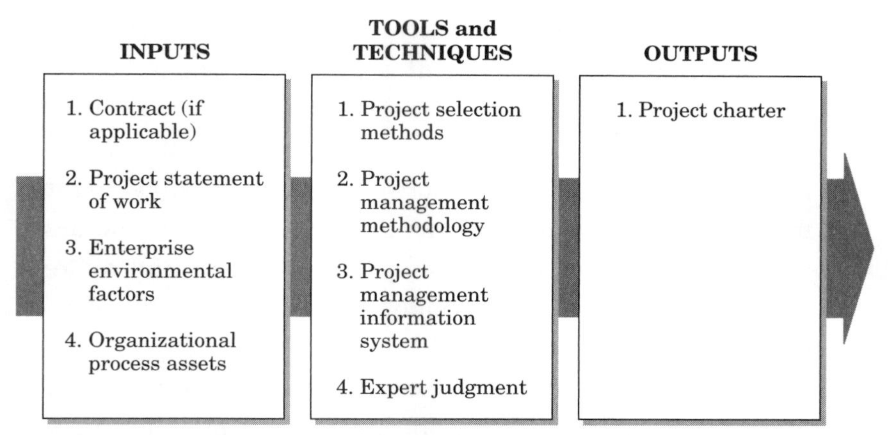

Figure 1.4 *PMBOK® Guide* 2004 Edition, Page 82.

input, how it was developed, and why it is needed. To further clarify, let's examine the inputs listed in Figure 1.4:

Contract. This is a legal document, the terms of which have been or will be agreed upon by the buyer and seller involved in a project. The contract may be an input in the project chartering process if the work to be done involves a supplier or vendor. The terms and conditions of the contract must be considered when making decisions about whether to approve a project. The risks, costs, and general terms and conditions should be reviewed in detail by the project manager and decision makers.

Project statement of work. The statement of work generally describes the specific work that will be done by a contractor. This information is also necessary when making project approval decisions.

Enterprise environmental factors. These factors are critical in the decision process and should be considered when determining which projects should be approved.

Enterprise environmental factors include the culture of the organization, government regulations and accepted business standards, the infrastructure and capabilities of the organization and many other factors. Any organization that is adopting formal project management methodologies should analyze the environmental factors that are specific to its company and industry and make sure these factors are considered throughout the planning and implementation of a project.

Organizational process assets. As organizations grow and mature, they will develop processes for doing business. These processes become part of the general operation for the organization and, in most cases, must be followed within the project environment as well. These process assets include safety procedures, purchasing processes, financial controls, and change control procedures. The project manager and team must be familiar with these processes to ensure compliance with internal rules and regulations and to prevent improper behavior that could result in injury, financial loss, or unnecessary expense.

Tools and Techniques

Project managers should possess some knowledge about the many tools and techniques that can be used to work with the inputs and process them to produce the desired outputs. The tools and techniques describe the necessary actions the team members or project manager will take and the specific items (the mechanism or device) needed to effectively process the inputs and produce a deliverable:

Project selection methods. These are the methods that may be used to determine which projects should be approved and authorized for funding and use of organizational resources. Project selection methods include techniques such as determining payback period of an investment, the present value of a future investment, net present value, internal rate of return, and break-even analysis. The use of decision trees and other forms of mathematical analysis may be used to determine which projects will be approved by the organization. Project selection methods analyze the tangible as well as intangible benefits of a project.

Project management methodology. The use of an approved methodology (a particular procedure or set of procedures) will generally promote consistency and efficiency. Organizations that manage projects on a regular basis and as part of their business when dealing with clients will develop a project methodology and promote its use throughout the organization. Sometimes, methodologies are developed by reviewing the processes and steps taken by project managers and teams on previous projects.

Project management information system. This is a system of systems working together and used by the project manager and project team to gather, store, analyze, and disseminate information about the project. These systems may include time reporting systems, financial tracking systems, project management software, resource management systems, and any others that may be used to manage information.

Expert judgment. There are many ways to describe the reference to expert judgment. The experts, generally the functional managers and technical experts, are the subject matter experts that are consulted or included in the project team and provide suggestions, guidance, estimates, and insights about risk, resource productivity, skills levels, and many other areas of importance.

Output

An *output* is the product of processing the inputs through the use of the tools and techniques. Outputs are deliverables and in many cases become inputs to several other processes. In this case, the output is a project charter:

Project charter. The charter is generally recognized as the document that authorizes the project to proceed, assigns the project manager, and begins the use of organizational resources. The project charter may be defined differently across industries. Any organization utilizing a project management methodology should have some type of document and/or process that clearly shows that a project has been authorized. An executive signature on a project charter or the project approval document can be a major influencing factor when it comes to obtaining resources and additional funding during negotiations.

As you can see, the *PMBOK® Guide* provides the basis for planning and presents a logical approach to produce an output or deliverable. In this example, using the process *Develop Project Charter* from the Integration Management chapter of the *PMBOK® Guide*, each input, tool and technique, and output is described in specific terms and from a project manager's perspective. In succeeding chapters the processes, tools and techniques, and other items that will help to accelerate planning will be explained in greater detail and through templates that are designed to efficiently expedite how a project is planned and managed.

THE *PMBOK® GUIDE*: LIFE FORCE FOR PROJECTS

The *PMBOK® Guide* can be used to assist in developing project plans and methodologies, but it is important to remember that it is a *framework* from which more detailed and customized project plans can be developed. If you try to use the *PMBOK® Guide* as the only means to manage a project, you will experience many challenges in your attempt to achieve project success. The *PMBOK® Guide* is arranged by knowledge areas and includes an explanation of the processes within each knowledge area. The knowledge areas are reviewed separately for explanation and learning purposes only. We don't manage projects by planning scope, time, cost, quality, risk, procurement, communications, and human resources separately. We all know that many planning processes are conducted concurrently, and there is a great deal of overlap in the processes we use. Using the combined information provided in the *PMBOK® Guide* along with some personal experience, logic, common sense and a touch of innovation—provides the basic formula for success. Let's call it "PMBOK®-based success."

You want your project to "live," to be seen for its value, to add to organizational effectiveness and the results to be used by the intended customers. If you think about it, your mission as a project manager is to bring your project to life, to obtain enthusiasm and commitment from your project team, and to achieve a feeling of accomplishment from your organization and gasps of awe from your customers. The *PMBOK® Guide* doesn't provide the excitement and drama experienced by novels such as Mary Shelley's Frankenstein. Nevertheless, your goal is to bring your project to life, just as Dr. Frankenstein proclaimed his project to be "*alive!*" His elation as he made this proclamation was, in a way, an indication of project success (or was it?). We certainly don't want our projects to result in mayhem, fear, and a generally unhappy stakeholder group (the townspeople, in the novel). We are looking for success from the customer point of view—and team member satisfaction, as well. The *PMBOK® Guide*, if used properly and with the appropriate techniques applied, will provide a foundation for success and can bring your project to life in a logical and effective manner and bring about a true sense of accomplishment for all stakeholders.

THE *PMBOK® GUIDE* PROJECT PLAN ACCELERATOR (PPA)

The *PMBOK® Guide* provides a description and brief explanation of numerous tools techniques, and processes that may be utilized to develop a project methodology or to improve existing organizational processes intended to guide project teams and project leaders. This Project Plan Accelerator can be used to assist you in developing best practices for your organization by elaborating in more detail on the elements of the *PMBOK® Guide*. Using the *PMBOK® Guide* as a reference, you can create customized processes, identify essential tools and techniques your organization would benefit from using, and create templates that will improve overall project performance. The *PMBOK® Guide* is an excellent source for developing customized processes and methodologies that will bring consistency to the planning and execution of projects within an organization. The PPA basically connects the higher level, broader project management body of knowledge with the *PMBOK® Guide* and can easily become the foundation for creating processes that are specific and customized to an organization but relate to the standards described in the *PMBOK® Guide*.

How to Use the Project Plan Accelerator

The Project Plan Accelerator will create a bridge between the *PMBOK® Guide*, the greater project management body of knowledge, and an actual project plan. Each item (input, tool or technique, output, process, etc.) discussed in the *PMBOK® Guide* that is targeted to be included in a project plan can be analyzed and further developed to ensure that the item is clearly documented, explained, and fully understood by the project team and other stakeholders.

Answering the Question: Why Do Project Managers Need This Book?

Here are some questions to consider for plan development and management of the project life cycle:

- Which elements of the *PMBOK® Guide* can be used to assist in developing a project plan or a subsidiary project plan for your project? Each chapter contains process information, descriptions of tools and techniques, definitions, lists, inputs, and outputs that could become key components of a project plan. Determine which elements, process groups, tool and techniques are most appropriate for your project.
- Which elements require additional research to ensure complete understanding and proper utilization?
- What tools or techniques are not applicable to the project? Why?

Consider the five process groups and the integration of the nine knowledge areas when developing your project plans and other project documentation. The emphasis should be on the relationships and interdependencies of the nine knowledge areas and how the five processes—initiating, planning, executing, monitoring and controlling, and closing—bring the knowledge areas together to form a systems approach to managing your project. Basically, every item in the *PMBOK® Guide* can be analyzed for inclusion in a project plan. Take any section of the *PMBOK® Guide* and select an item. As an example, let's use the initiating process group:

Initiating process group. Determine the specific actions or processes that should be applied to ensure that projects and project phases are authorized to proceed. Think of your current project work environment. Think of the enterprisewide factors that will influence these decisions. Identify the organization's processes that must be followed by the project team. Who are the key stakeholders that will make the decisions? What are the benefits associated with this project? Is there a business case with detailed information about the anticipated results? If the project is in progress, is it worth continuing to the next phase? Are the risks acceptable? Do you have enough funding? Are the variances acceptable?

For each component that has been selected, an analysis is completed to ensure that all major issues and questions have been identified. Once the questions have been asked or the issues raised, the next step to take action to resolve the issue. In some cases, additional research is required before a decision can be made. This process can be scaled to meet the specific needs of the project. We know that each project is unique, and it may not be necessary to use all of the processes, tools, and techniques that are available. However, using an approach like this actually helps to accelerate the planning process by organizing the elements that should be considered, analyzing each element to the appropriate level of detail and documenting the findings for reference later. This process will also create a very useful lessons-learned file and may help to improve many existing organizational processes. Think about projects you are currently engaged in. Where are the obstacles?

What project components or planning issues require additional detail or research? Make a note of these items for reference later.

Notes: ____

____

____

____

____

____

____

____

____

____

____

____

The Project Plan Accelerator can effectively identify the main issues a project manager and team may encounter during project initiation and through project execution and closure. As you can see in the example, the components necessary for project planning have been identified and listed. The component is then analyzed, actions are assigned, and the supporting detail is obtained. The complexity of the project will determine how much detail is needed to develop a solid, workable plan. The byproduct of the Project Plan Accelerator is that it will become a lessons-learned library or a series of supporting documents that will further improve the efficiency of the planning process.

PMBOK® Guide Component / Tool / Technique / Process	*Analysis, Implementation Activities, and Research Required*	*Person Responsible*	*Date Required*
Example: Enterprise environmental factors	Define the factors within the organization's environment that will have an impact on the project planning process. Example: What departments or organizations will be involved in the project? What cross-cultural issues must be considered (within the organization and external to the organization)? What government or industry standards or regulations must be considered during the planning process? How well will the organization's infrastructure support the project? What is the organization's tolerance for risk? How does this tolerance affect project planning?	Identify specific team members or stakeholders who will research the required information. Set expectations about how the information will be prepared and delivered.	
Example: Organizational process assets	Identify the processes that exist within the organization that must be followed—Include business related processes that may affect planning. What safety procedures must be in place for the project?		

	What quality assurance processes are currently in use and must be included in the project planning process? What is the organization's Quality Policy? Are project planning templates available? Where are they filed? What financial controls are in place within the organization for time reporting, tracking costs, progress payments, and allocation of funds?		
Example: The Project management office (PMO)	Define the current responsibilities of your organization's PMO. What processes have been defined? What training is available to project managers? What is the level of support provided by the PMO? What additional benefits may be gained by expanding (or creating) a PMO? What key measures of success have been identified by the PMO or should be tracked by the PMO?		
Example: Project constraints	What are the specific constraints associated with the project (schedule, resource availability, funding, skill levels, contractual items etc.)?		

(continued)

Project Plan Accelerator (*Continued*)

PMBOK® Guide Component / Tool / Technique / Process	*Analysis, Implementation Activities, and Research Required*	*Person Responsible*	*Date Required*
Example: Project management information system	What automated systems will be used to manage project information (project software, enterprise-level software, time reporting systems, financial reporting systems, document management systems, repositories)?		
Add components as needed and complete the columns and rows by developing and then answering questions about the project and the plan using the terminology described in the PMBOK® Guide.	Ask yourself why this planning component is important, and determine how much you should know about it. It is necessary? Why? How will you use it? Do you fully understand what it is, as well as its overall importance? Remember: The *PMBOK® Guide* provides information that *could* be used in project planning but some componenets may not be necessary for your project.		

Chapter Two

The Big Picture

Project management has been practiced since the days of the pyramids and project methodologies have been evolving for decades. This evolution includes everything from resource management to network diagram development to leadership and motivation techniques. I like to think that project management has evolved into an essential element of overall business management and includes a very unique and desirable skill set. I see it as a field where leadership and management skills become balanced and provide an essential element for overall organizational success not only at the project level but at the strategic level as well. Project managers are trained to look at the larger picture and focus on things such as vision, objectives, goals, coordination, and integration. When the *PMBOK® Guide* was originally developed, the main focus was on eight knowledge areas:

1. **Scope management**. What must be done?
2. **Time management**. When should it be done?
3. **Cost management**. How much will it cost?
4. **Quality management**. How good should it be?
5. **Human resource management**. Who will do the work?

6. **Communications management**. How will information be delivered?
7. **Risk management.** What problems may be encountered?
8. **Procurement management**. What material, supplies, talent, and equipment must be obtained?

The ninth knowledge area, Integration management was added in the 1996 edition of the *PMBOK® Guide*. These knowledge areas were defined in terms of their specific lower-level components and process groups and were described independently for learning purposes. The connection between scope and time management or human resource and quality management was not clearly defined, although the areas were very much linked in the overall planning processes. As we look at how projects are managed today, we can see that we do not focus first on scope, then on time management, then on risk management, and so on. Project managers consider the linkages between each area of knowledge and create integrated plans with the input and expertise of the project team members (the representatives of the functional groups) and other key project stakeholders. This integrated approach ensures the development of a more complete project plan.

In the ideal project management world, the project manager would be given the opportunity to review, select and arrange to have the appropriate resources assigned to a project and would then manage the planning process from a facilitator/leader perspective. The project manager would not be the technical expert, but would have a sufficient understanding of the technology involved to coordinate activities, discuss issues with clients and subject matter experts, manage interfaces with functional groups, and make decisions regarding project performance or changes in the project plan. This is what can be referred to as the *big picture view*. In actual project practice, the project manager is often required to manage the overall project, and will also be responsible for a portion of the technical work and some deliverables. In these cases, the project manager must be able to shift from a managerial role to a technical role and back again. This presents somewhat of a challenge to the project manager as there will always be a need to observe the project from a higher-level or cross organizational viewpoint. An understanding of how projects are managed from start to finish, how the project may impact an organization's daily operations, and how all of the components of the project are integrated is essential for success. If we examine the relationship between the knowledge areas and the five process groups, we can see how the overall project management process comes together.

The five processes (Initiating, Planning, Executing, Monitoring and Controlling, and Closing), which contain subsets of lower-level processes associated with the nine knowledge areas, can be used to manage a specific project phase or the entire project life cycle. Identifying the subprocesses, activities, and deliverables associated with each major process group will provide project teams and stakeholders with a fairly complete picture of what is required to plan and complete a project successfully. This approach can be used by any organization to develop a universal project methodology and can be customized, changed, and updated relatively easily. Relating the five process groups to each phase of the project as well as at the project life cycle level provides a greater foundation for more effective planning and overall project control.

Figure 2.1 shows how the five process groups integrate the subprocesses associated with each knowledge area to create a basic flow of how a project is managed from start to finish. This higher-level process can be customized to meet the specific project planning requirements of any organization. These are not phases of a project but a logical flow of the work that

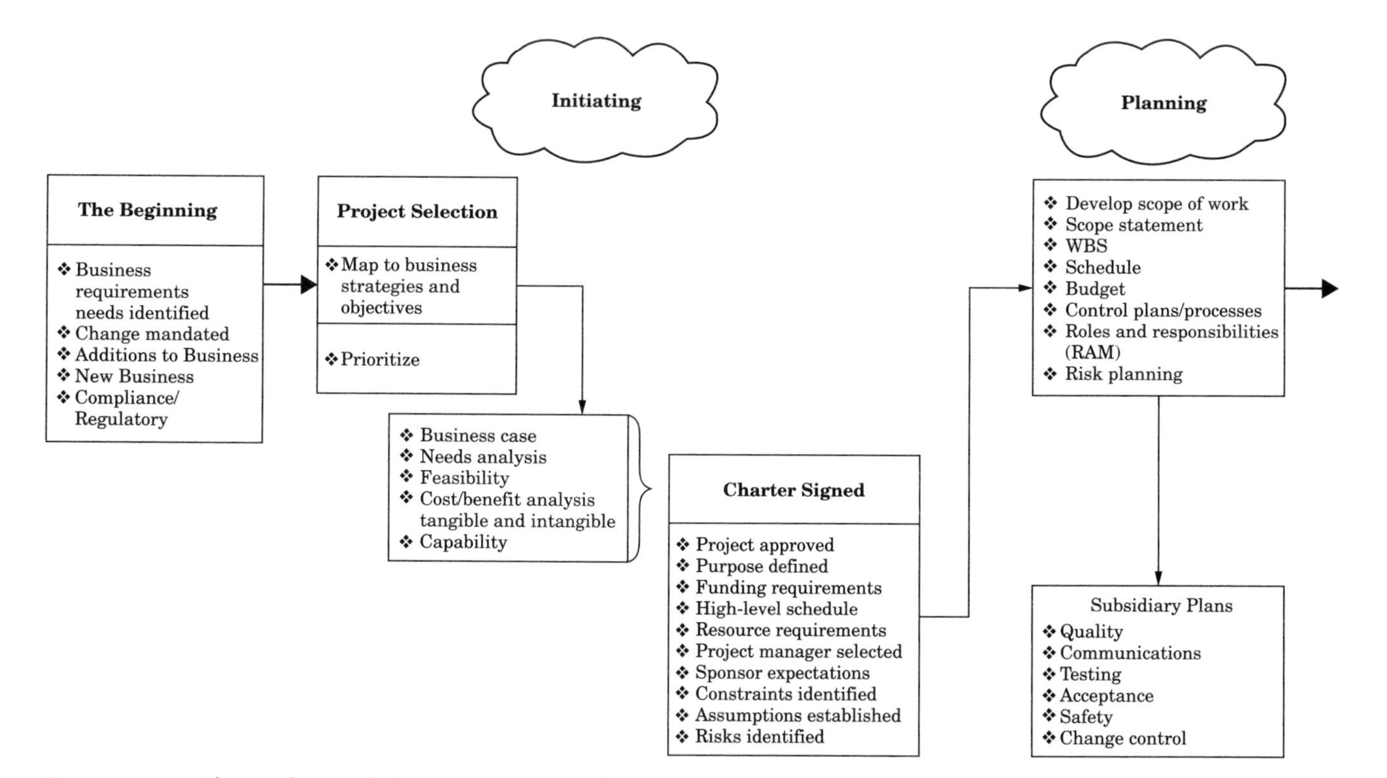

Figure 2.1 Flow Chart of the Typical Activities and Deliverables Associated with a Project Management Process.

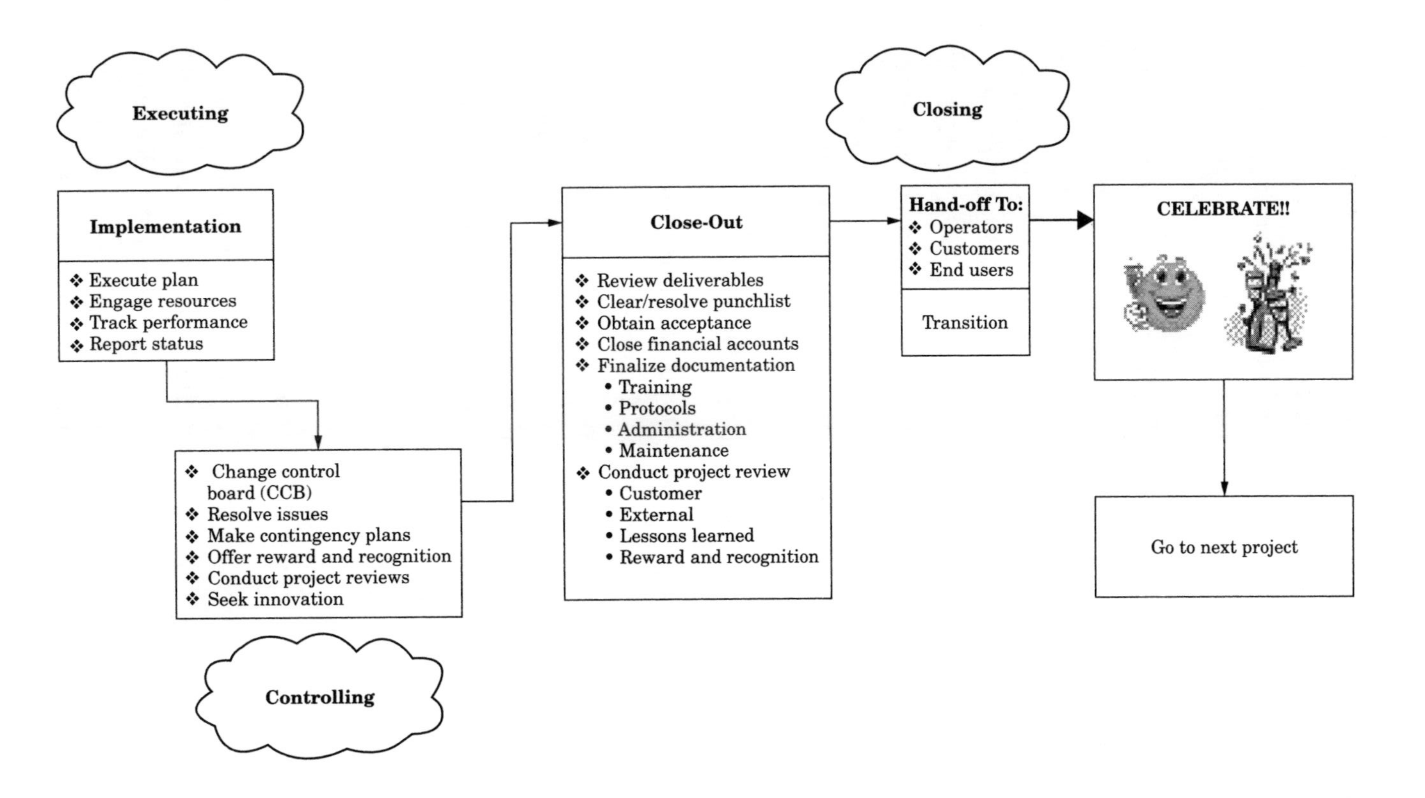

Figure 2.1 (*Continued*)

would be performed to complete a project. The *PMBOK® Guide* associates the five process groups with the sub processes found in each knowledge area to create a more integrated planning process where many planning activities are actually implemented concurrently.

Using the flowchart as a basis for discussion, a project manager can clearly explain to a project team or to stakeholders who are not familiar with formal project management procedures how a project is planned, executed, and eventually closed. This flowchart is the basis for *Bringing the PMBOK® Guide to Life*. The intention is not to compete or replace the *PMBOK® Guide* but to provide an enhancement and a convenient reference and support resource that is especially useful for organizations where formal project management may be resisted.

THE INITIATING PROCESS

All projects go through some type of initiating process. Sometimes the process is very quick, especially for small, minimally complex, and short-duration project. In other cases, the process can become tedious, lengthy, and challenging to the people involved. Regardless of the type of project, even if the project involves moving a person from one office to another, replacing a computer, or painting a room in your home, some type of reasoning and validation for the project is required.

The Beginning

Most projects begin with some type of need or problem to be solved. Things to consider during the process of initiating a project include the business objectives, business requirements (what would be needed to achieve the business objectives), a change that is mandated by management and may require significant effort to accomplish, additions to the business such as new products or services, creating a new business division and a new product line, or compliance with a government regulation. The initiating process should include some dialog about why a project may be necessary.

Project Selection

During this part of the process, projects that are being considered are analyzed for *value* to the organization. Project selection may include several different techniques, and the rationale for selection may also vary. Selection includes analysis of several factors:

- Is there a business case for the project? How well does the business case explain the project?
- Does the project map to the organization's overall strategy and long-term plan?
- What are the organizational limitations—available funding, resources, current workload?
- Does the organization have the capability to perform the work? Is it feasible?
- What is the benefit of performing the project in terms of the tangible (example: return on investment) and intangible benefits (affect on reputation, employee morale, social acceptance)?
- What are the specific capabilities required and does the organization have those capabilities? Examples: skills, machinery and capacity of production groups, expertise in the product or service.

As projects are selected and included in a project portfolio, the organization should arrange projects in order of priority. Organizations generally do not have unlimited resources, and there will be occasional decisions to sacrifice a less-important project for a more critical project in terms of use of human and other resources.

Project Charter Signed

The project charter is generally defined and accepted as the document that authorizes the project to use organizational resources. The charter indicates that the project has gone through the selection process and that a sponsor sees the value of the project, or there is agreement that the project is necessary. The purpose of the project is defined, a high-level estimate of the project budget is prepared, there is some information about the schedule, mainly in terms of milestones, the resources and various functional groups are identified, and the project manager is assigned. (Resource requirements and other factors will change as the project is further defined.) The project charter may also define the specific expectations of the project sponsor and the level of authority of the project manager. An initial set of project assumptions and project constraints may also be included in the document.

Sample Outline of a Project Charter

There are many examples of project charters available for reference. In some organizations the project charter, statement of work, and scope statement are included in one document. The charter is not a project plan but becomes part of the planning process. Typical components of a project charter include:

Project Name and unique identifier. Some organizations assign specific project numbers for ease of tracking.

Business need. Why the project has been approved.

Product or service description. What the project will produce in the form of deliverables.

Stakeholder identification. The organizations and individuals that will be assigned to the project or affected by the project. This includes business units, individuals, functional groups, customers, contractors, and people negatively affected by the project results.

Summary schedule. Milestones and contractual dates.

Assumptions. Planning items that are based on an assessment of the organization's environment, resource availability, organizational capabilities, and other factors.

Constraints. Limitations that must be considered during the planning and execution of the project.

Budget. High-level estimate (rough order of magnitude) of the cost of the project. The budget will be defined in greater detail as the planning process is engaged.

*Project Charter Table of Contents**

A project charter may appear in the form of a detailed document that includes a well-organized table of contents. This depends on the organization and the level of maturity of the project management methodology. The charter document may include the following items:

Section 1. Project Overview
- 1.1 Problem Statement
- 1.2 Project Description
- 1.3 Project Goals and Objectives
- 1.4 Project Scope
- 1.5 Critical Success Factors
- 1.6 Assumptions
- 1.7 Constraints

---

*Sample from Texas Department of Resources, www.dir.state.tx.us/pubs/framework/gate1/projectcharter.

1.8 Assumptions
1.9 Constraints
Section 2. Project Authority and Milestones
2.1 Funding Authority
2.2 Project Oversight Authority
2.3 Major Project Milestones
Section 3. Project Organization
3.1 Project Structure
3.2 Roles and Responsibilities
3.3 Project Facilities and Resources
Section 4. Points of Contact
Section 5. Glossary
Section 6. Revision History
Section 7. Appendices

This example clearly shows the major elements associated with a project charter and can be used as a model for developing a customized charter document for many organizations.

THE PLANNING PROCESS

The planning process includes elements from several project management knowledge areas. The process is based on an integration of these elements to develop a well-organized and complete project plan. During the planning process, the project scope is defined in more detail (this is referred to as the scope definition process). A **project scope statement** is created and defines the project in much more detail than the project charter. The general understanding of a scope statement is that it answers the following questions about the project: Who is involved? Why it is needed? When must it be done? Where will it be done? How will it be done? How much will it cost? How many deliverables must be produced? Who will do the work?

The scope statement is not a plan but it is a document that will help the project team develop a comprehensive plan by basically providing enough detail for the team and other stakeholders to fully understand what the project is all about. During the planning process, the project team and project manager will use several tools and techniques and spend a significant amount of time asking questions, gathering data, reviewing lessons learned, and revising the scope of work to ensure that all stakeholders fully understand what must be accomplished. Typically, the planning process includes activities associated with the following items:

- *Scope statement development.* The scope is defined in detail. In some cases the scope statement is derived from the project contract or statement of work.
- *Work breakdown structure (WBS).* Decomposition of the project in steps and layers to display the project in increasing levels of detail.
- *Work Breakdown Structure Dictionary.* This document is sometimes used to provide stakeholders with additional information and details about project tasks and activities
- *Project schedule.* The arrangement of project activities in predecessor/successor relationships to create a network diagram that shows the logical flow of work and the estimated duration of project activities to indicate the total duration of the project. The project's critical tasks are also defined. The project schedule is the result of a coordinated effort by the project manager and the team.
- *Project budget.* The total estimated costs of the project, including human resources, equipment, tools, materials, and other costs such as travel, rent, supplies, conference facilities, and recurring project costs.
- *Control plans and processes.* These are the processes that will help the project team to control changes, identify

variances to the plan, escalate issues, and identify project jeopardy situations. Control procedures should be well documented, mutually agreed upon by the stakeholders, and enforced to minimize project disruptions and unauthorized changes. Without controlling processes, the project team may experience a common problem known as **scope creep**—the uncontrolled addition of work into the project.

- *Roles and responsibilities.* The planning process includes the identification of key project stakeholders and the ability to align these stakeholders with the project tasks identified during the WBS development. Clearly defined roles and responsibilities will reduce conflicts, promote more efficient use of resources, and help to develop a high-performing team. A common approach and tool for defining roles and responsibilities is the **responsibility assignment matrix (RAM)** that aligns the major project tasks identified in the WBS with the functional managers assigned to the project. This is shown in Table 2.1.
- *Risk planning.* During the planning process, the project stakeholders will assess potential threats and opportunities, identify potential risk events, and plan strategies to respond to identified risks. Risk responses are associated with the following actions:
  - *Avoidance.* Attempting to find an alternative to a risk situation that is unacceptable and potentially dangerous or could have severe consequences.
  - *Mitigation.* Attempting to reduce the probability and impact of the identified risk.
  - *Transfer.* Identifying a resource that can manage the risk more effectively and handing over the risk situation to that resource. This is not avoidance of the risk; the risk has been identified and a qualified resource or

Table 2.1 Responsibility Assignment Matrix

WBS Element	*Team Member Name (Project manager)*	*Team Member Name (Client)*	*Team Member Name (Engineer)*	*Team Member Name (Quality Assurance Mgr)*
1.1	R	I	C	C
1.2	C	I	R	C
2.1	C	R	I	C

R = Responsible—This identifies the person responsible for the completion of the task or WBS element
I = Inform—This person must be informed of task status
C = Contribute—This person is involved in the delivery of the WBS element and may provide resources or effort to assist in the completion of the element.

organization will assume the responsibility of managing the risk.

- *Acceptance*. The acceptance of a risk indicates that the project team understands the nature of the risk situation and has either developed a plan to manage the risk situation if it occurs or it is confident that the risk will be handled effectively if it does occur. Additionally, there are responses for managing risks that are considered to have a positive effect on the project: Exploiting a risk situation—Actually attempting to encourage the situation to happen due to its potential positive impact; Enhancing—working to obtain the greatest benefit of the situation; Sharing—advising others in the organization about the positive impact of the risk event.
- It is important to note that project risk management may also be associated with opportunities or positive

outcomes. In situations where opportunity exists the project manager and team should focus on exploiting the opportunity (making it happen), enhancing the opportunity (maximizing the benefits) and sharing the opportunity (improving overall organizational performance)

- *Subsidiary plans*. Depending on the type of project and its complexity, the project team may develop supporting plans for the overall project management plan. These plans provide guidance about control procedures and decision making and a more detailed level of information for elements of the project plan that are generally directly connected to a specific knowledge area of the *PMBOK® Guide*. Subsidiary plans are developed based on the needs of the project.
- Examples of subsidiary plans:
  - Scope management plan
  - Cost management plan
  - Risk management plan
  - Quality management plan
  - Communications management plan
  - Procurement management plan
  - Schedule management plan
  - Human resource management plan
  - Safety management plan
  - Change control and configuration management plans
  - Testing and acceptance plan
  - Cutover and handoff plans
  - Reward and recognition plan
  - Project close-out plans

Subsidiary plans will vary, depending on the project type, the methodology being used, the culture or the organization, the managerial styles of the organization and the urgency associated with the project. The detail of each of these plans depends on the project type and the project management maturity level of the organization.

EXECUTING, MONITORING AND CONTROLLING, AND CLOSING PROCESSES

Executing

At the completion of the planning process, the project manager and team will begin to execute the tasks and activities that have been planned. The assigned resources are obtained, engaged in the project and begin to perform their specific project activities. Procurement activities such as RFP submissions and seller selection begin and contracts are negotiated. As work is performed, results will be produced that must be analyzed. The project manager obtains status from the project team and tracks performance against the plan. Status reports are generated as described in the communications plan (a subsidiary plan defining the process in which progress, status, and forecast information is developed and distributed) and the project manager uses techniques such as *earned value analysis* to determine where variances are occurring.

Changes are introduced by project stakeholders during execution and are managed through the change control plan developed in the planning processes. Quality assurance managers assess the work and compare it to standards and defined requirements. Issues are raised by the project stakeholders and are analyzed for impact on the project, and solutions are developed. As risk situations are encountered, the project team utilizes contingency plans or develops responses to manage the issue. During execution of the plan, the project manager should exercise a reward and recognition process and actively acknowledge team members for excellent work and the achievement of milestones. This regular, and sincere, acknowledgment will help to maintain team morale and loyalty.

Project reviews are conducted throughout the project life cycle. These reviews help the team to determine what additional actions may be necessary to correct a variance or determine if the project should continue into the next phase. During execution, project teams may be required to innovate and use creativity to resolve issues, especially in an environment where funding, time, and resources are extremely limited. While the project is being executed the project manager and the team will engage in monitoring and control activities that will identify where variances are unacceptable. The use of control charts, Pareto diagrams, root cause analysis, change control systems, configuration management processes, and earned value analysis will provide the data to minimize risks and identify symptoms that could result is serious project problems in the form of schedule delays, cost overruns or safety issues.

Closing Process and Project Closeout

The closing of the project is a critical part of the overall project life cycle. The project team, as part of the planning process, should have developed a plan for closing out the project. There is a very specific and important need to determine what is required to actually achieve acceptance of the project. During the closing process, the project deliverables are reviewed to ensure compliance with the scope statement and the project contract. Often, the project team will be provided with a **punch list** (a list of unacceptable items that must be corrected) from the client. A carefully developed plan and a strong and well-executed monitoring and control process will help to minimize the size of the punch list.

The objective of the closing process is to obtain formal acceptance from the client or intended user of the product or service. Formal acceptance criteria are developed in the planning process as a subsidiary plan (as needed).

As the project progresses through the close-out process, any open financial accounts should be updated and then closed as soon as possible to prevent additional charges to the accounts. The team reviews all project documentation and prepares all required financial statements, training procedures, administrative processes, protocols, maintenance procedures, and any other items identified in the contract or other project agreements.

THE PROJECT REVIEW

The project plan should include the scheduling of project reviews throughout the lifecycle of the project. Ideally, a project review should be conducted at the end of each phase. The review process will assist in identifying best practices, as well as areas where additional improvement is needed. Project reviews may be completed in stages. Customer reviews are scheduled to obtain information from the user's perspective, and internal team reviews may be scheduled to discuss what the team accomplished well and where the team should focus attention for improvement. During the review process, the team identifies lessons learned and improved ways of performing work. These lessons learned may be used to update the organization's processes and procedures and improve the capability of the organization to manage projects successfully.

A WORD ABOUT RECOGNITION AND ACKNOWLEDGMENT

The closing procedure should include some type of reward and recognition event. The type of project, the available funding, and other factors must be considered, but regardless of project size, recognition of work well done is critical and an absolute necessity. People do like to receive an acknowledgment occasionally, and lack of acknowledgment has many negative side affects. It is important to note that culture will be a factor to consider when determining an appropriate recognition process. Project managers should, whenever possible, include in the project plan a budget for recognition. If funding is not available or organizational procedures do not create an opportunity for formal recognition, the project manager should at least take the time to recognize the team for the success through letters, certificates of appreciation, or other activities. A well delivered and sincere acknowledgment will generally pay off in terms of greater loyalty, strengthened relationships, and a willingness to step up to the next project with confidence.

THE *PMBOK® GUIDE* PROJECT PLAN ACCELERATOR (PPA)

Using information from this chapter and the *PMBOK® Guide* for additional reference, make a note of any specific planning items that should become components of your project plan. Describe the component and develop questions around the component that will enable you to effectively use the component in the plan.

The Big Picture

Developing a high-level Project Plan Accelerator document requires the project manager and project team to view the project from an "executive" or strategic viewpoint. To develop the big picture, you might ask yourself these questions:

- Which elements of the overall planning process will be used in the management of the project? Use the *PMBOK® Guide* as your reference source.
- How complex is the project? What do you know about it?

- Who are the stakeholders and what are their respective roles?
- What planning documents are available within the organization?
- What lessons learned and best practices have been documented?
- What training may be needed for the functional managers and team members?

Consider how the five process groups and the knowledge areas will be used during planning and execution. Determine what is missing from your process by analyzing your organization's internal processes and lessons learned files. Remember that planning is an iterative process and plans will change as more project details emerge.

Notes:____
____
____
____
____
____
____
____
____
____
____
____

Project Plan Accelerator

PMBOK® Guide Component / Tool / Technique / Process	*Action Plan for Implementation (Develop questions that could be researched by team members.)*	*Person Responsible*	*Date Required*
Example: The beginning	How do projects begin in your organization? Refer to the flowchart for ideas. What actions are typically taken as a project begins? Who is involved and what are they responsible for? List these steps or activities.		
Example: Initiating	What are the steps used to initiate a project in your organization? This would include a review of project selection criteria. (What selection criteria is used?) How is the project formally authorized? What are the business needs and requirements the project will address? What are the project objectives? What is the connection between the project and the strategic plan? Is a project charter available? Has a preliminary scope statement been prepared?		

Example: Planning	Outline your planning process. What can be added or omitted? What are the project risks? What project dependencies have been identified? What opportunities exist? What are the project constraints? What assumptions have been identified? How will cost and activity duration be estimated?		
Example: Executing	How will the organizations involved receive direction? How will the project team be acquired? Who will perform quality assurance? How will project information be distributed?		
Example: Monitoring and Controlling	What control procedures have been developed? Is an integrated change control process in place? How will deliverables be reviewed and accepted? (Scope verification.) What is the process for managing scope, cost, and schedule changes? What is the process for managing contract administration?		

(continued)

Project Plan Accelerator (*Continued*)

PMBOK® Guide Component / Tool / Technique / Process	*Action Plan for Implementation*	*Person Responsible*	*Date Required*
Example: Closing	What is the process for closing out your projects? How effective are your project reviews? Do you have a defined acceptance process? How do you recognize your project team? What is the process to ensure compliance with contractual obligations? Is there a formal project closing process to finalize all activities?		

Chapter Three

Project Plan SWOT Analysis

"The first casualty in any battle is the plan."
—Napoleon

Project managers, especially those who work in organizations where risk management is a high priority or where strategic planning is emphasized, will be familiar with a **SWOT analysis** (strengths, weaknesses, opportunities, threats). With just a bit of imagination, the SWOT analysis technique can be applied to the development of a project plan. Managers view the **SWOT Analysis** as a strategic planning method used to evaluate the **S**trengths, **W**eaknesses, **O**pportunities, and **T**hreats involved in a project or in a business venture. It involves specifying the objective of the business venture or project and identifying the internal and external factors that are favorable and unfavorable to achieving that objective. From a project manager's perspective a SWOT analysis is also associated with risk planning. Regardless of viewpoint the technique is a useful planning technique. There is a quote most project managers are familiar with: "If you fail to plan, then you plan to fail." I also refer to a quote attributed to Napoleon: "The first casualty in any battle is the plan," and a quote by General Dwight D. Eisenhower: "In preparing for battle, I have always found that plans are useless, but planning is indispensable."

Clearly, planning is critical to success, and having a plan provides the basis for decision making, control and effectively managing the project. We go through a planning process to develop plans that will set the direction of the project and keep the project team on track. The approved plan is executed and, inevitably, change occurs. We can expect change, mainly because all plans are based on estimates (guesses) and knowledge or lessons learned from the past. The uncertainty of the project's future and the inability to actually predict a definitive outcome will require us to continually update and change plans to meet the desired objectives. With uncertainty in mind, we can use the SWOT analysis process to develop a plan framework, and eventually a project plan that addresses project risk at the appropriate level. A project plan SWOT analysis will not eliminate the need for changes or remove uncertainty, but it can reduce many unnecessary changes and minimize, to some extent, the probability that some important planning components will be omitted. Refer to Table 3.1 for an example of a plan SWOT analysis.

Table 3.1 Project Plan SWOT Analysis

Strengths	*Opportunities*
What specific components are included in the plan that will clearly assist the team in achieving the project objectives? Consider what can be done (if anything) to further strengthen the plan. Examples: Is a clearly defined Communication plan in place? Has the team developed an effective and agreed upon Change management plan?	What opportunities have been identified during the development of the plan or as a result of the completed plan (at the project level and the organizational level)? Examples: A more streamlined and efficient planning process can be documented from lessons learned.

Is a Risk management process in place and communicated to all stakeholders? Are Contingencies included in cost and schedule estimates? Is the plan concise and written in a format that will be understood by the stakeholders? Does The plan include specific measurable objectives? Is the administrative burden (extra paperwork, reports, excessive meeting requirements, approval processes) minimized?	The PMO will be able to develop a more consistent organizational planning process. The plan approval process can be accelerated. There will be greater levels of customer and key stakeholder satisfaction.
Weaknesses	*Threats*
What weaknesses can be detected during plan reviews? Examples: Language is unclear. Technical information is ambiguous. Input from functional units is incomplete. The plan does not address the full scope of the project. Information is not provided at the appropriate level of detail.	What threats exist that should be removed or mitigated? Examples: Pressure for action from management may result in incomplete plans. Lack of a clearly defined plan may result in loss of support from management. Lack of input from end users may result in significant resistance to the plan. Planning is not a priority of the organization.

A project plan SWOT analysis will assist the project manager in identifying where additional plan development work is required. The key is to enhance strengths, resolve and strengthen weaknesses, exploit opportunities, and remove or minimize threats. The action items developed to address these areas, when properly executed, will result in a strong, well documented, and more complete project plan.

PROJECT PLAN ACCELERATOR

Using information from this chapter and the *PMBOK® Guide* for additional reference, make a note of any specific planning items that should become components of your project plan. Describe the component and develop questions around the component that will enable you to effectively use the component in the plan.

Project Plan SWOT Analysis

- Discuss the swot analysis process with your team members. Explain the technique and the desired outcome.
- Provide examples of strengths, weaknesses, opportunities, and threats. Using examples to explain the process will accelerate the learning curve and produce a more complete and usable document.
- Use the Project Plan Accelerator template to identify the project plan components. (Refer to the five process groups and nine knowledge areas to further identify plan components.)

Consider how the five process groups and the knowledge areas can be used during the SWOT analysis. As an example: The planning process group includes scope planning, create the WBS, activity definition, activity resource estimating and cost estimating. The SWOT analysis can be used to determine if these items have been developed to the appropriate detail or if they have actually been included in the plan. Review the planning process in use by your organization. What are your planning strengths? How do you know they are your strengths? (Performance information provided by a PMO can assist in supporting identified strengths.) What are the performance metrics and measurements that support your observations? What are your weak areas? Why are they weak? What can be done to strengthen them? What opportunities are possible if the plan is well defined and the project is successful? If your plan is not well organized and complete what threats exist that can disrupt the planning process and the execution of the plan? What threats may develop that can impact the reputation and well being of the organization?

Notes:____

Project Plan Accelerator

PMBOK® Guide Component / Tool / Technique / Process	*Action Plan for Implementation*	*Person Responsible*	*Date Required*
SWOT analysis	Review the project plan at several stages during its development and conduct a SWOT analysis to determine where additional effort is required. Arrange to have the plan reviewed by a recognized, experienced planning expert. Create a SWOT analysis template and discuss each component of the analysis with the project team or other stakeholders.		
Example: Strengths	Plans are developed through the input of the project stakeholders. Assess the plan and identify where the plan is well written, clearly defined, and where risks have been minimized.		
Example: Weaknesses	Identify elements of the plan that are ambiguous or use complex sentences that are difficult to interpret or can lead to confusion or conflict.		

Example: Opportunities	Identify elements of the plan that can be communicated to other project teams and planning groups to improve overall organizational planning. Identify breakthrough thinking and streamlined techniques that may benefit the entire organization. Identify where organizational process assets have been improved and should be shared.		
Example: Threats	Identify poorly worded paragraphs that may be interpreted differently by project stakeholders. Identify ambiguous phrases, unclear or unfamiliar terminology. Determine if there is language that could be damaging to the organization or where unreasonable conditions have been defined.		

Chapter Four

Developing a Project Methodology

A methodology is defined as a system of practices, techniques, procedures, and rules used by those who work in a discipline. Establishing a methodology and ensuring its consistent use provides an organization with many benefits including consistency, greater predictability or outcomes, and better management of costs and resources. The *PMBOK® Guide* provides a basis for developing an organized approach to managing a project and has been used by many organizations to develop a methodology that works effectively for their specific enterprise. The processes and procedures described in the *PMBOK® Guide* have been developed and written by project managers from a wide range of industries, including nonprofit organizations and government agencies. They can be used to create a framework for managing projects in a consistent manner that is acceptable for use by all business units and work entities within an enterprise. The important thing to remember is that the *PMBOK® Guide* is a collection of best practices. The processes and procedures defined in the document many not work for every project and some modification will be required. Using common sense and appropriate application of tools and techniques is the most effective and preferred approach of the successful project manager.

PROJECT METHODOLOGY QUESTIONS

Before developing a project management methodology, it would be beneficial to analyze the current environment of the organization using a series of questions and a detailed needs analysis. The following is a list of twenty questions that may be considered during a project management methodology needs analysis:

1. What benefits can be gained by implementing a formal project management methodology within the organization?
2. How would you describe the current project management process or methodology? If there is no formal methodology, try to determine what processes seem to be common or repeated.
3. What is the greatest challenge encountered by the project manager during project planning and execution? There may be a few of these challenges, so make sure you include the ones that seem to be most prevalent.
4. What metrics or performance measures are currently being used to determine project performance levels? If there are no official metrics used, identify the metrics you use to determine performance. These metrics may include: schedule progress, completed deliverables, activities that have begun, activities in progress, quality (number of defects or repairs), actual cost, percent complete of an activity, efficiency in the use of resources.
5. What metrics are missing and should be included in the process of managing project performance? Compare with other project-based organizations.

6. How would you describe the efficiency of the project status meetings that are scheduled and conducted within your organization? How well are your meetings managed? What information is included in project status reports?
7. How does a project team identify all of the project stakeholders? Do project team members understand the true definition of a stakeholder?
8. How are project stakeholder requirements determined and documented? This includes requirements about information distribution.
9. How often is project status communicated to project stakeholders? What criteria are used to determine who should receive information and when?
10. How are projects selected for implementation? What is the specific selection process used for determining which projects are most beneficial for the organization?
11. What is the process for reviewing/auditing project performance?
12. What formal project management training is available for the organization? What percentage of the organization has attended some form of project training?
13. How are changes to project plans and project baselines managed? What are the organization's policies or processes for managing change?
14. How is a project team selected?
15. What is the role of the project manager?
16. How are project risks identified and managed?
17. What criteria are being used to determine project success?
18. When is a project considered to be complete?
19. What is the process for closing out a completed project?
20. What is the process for project reviews?

A review of the Big Picture flowchart described in Chapter Two (see Figure 2.1) and the use of the *PMBOK® Guide* will provide the essential information needed to develop a project management methodology. Generally, the development of a project management methodology is associated with a project management office (PMO). Lessons learned from the development of PMOs in the mid–to–late 1990s indicate that the introduction of a project management methodology should be gradual, with emphasis on engaging the managers and executives of the organization. Explanations about the benefits of project management—such as consistency of delivery, higher quality, and more efficient use of resources—should be provided through studies, reports, and presentation by enthusiastic promoters. Attempting to introduce a project methodology through force and by using and demanding extensive administrative tools and procedures will generally result in resistance.

Caution should be taken to avoid introducing too much in the form of formal procedures and standard forms, which may cause a backlash from a perceived increase in administrative work. A project methodology should be designed with the people who will use it in mind. I and many other projects managers have heard a familiar cry, "Don't make a project out of it!" This statement of anguish is an indication that project management is not universally accepted as a key success factor for business and must be introduced through a series of predetermined steps and with significant emphasis on the successes achieved along the way.

The project management office is generally the main driver of project methodologies within an organization. Figure 4.1 illustrates the many elements required to develop a PMO, create accepted methods, and provide opportunities for continued growth. Remember, caution should be exercised when attempting to create a methodology and a project office. Additional lessons learned from the mid–to–late 1990s indicate that an aggressive approach to developing a project management process or creating a Project Management Office can result in significant resistance by the project managers, project teams, and managers or executives whose resources may be affected by the introduction of new methods and procedures. An approach that appears to work effectively is to schedule focus groups and needs analysis sessions to identify the concerns of the various business entities and functional groups that may be affected by the new methodology. Communication is essential for any methodology to be developed and accepted by the organization. The purpose of introducing a project methodology within an organization is to improve efficiency, not to create resistance. Patience is the key here. Take the time to create a plan for introducing or rolling out a new methodology. Obtain feedback and comments from the people who will be affected. Maintain a message of "continuous improvement" and focus on gaining support through logic and with the organization's well being in mind.

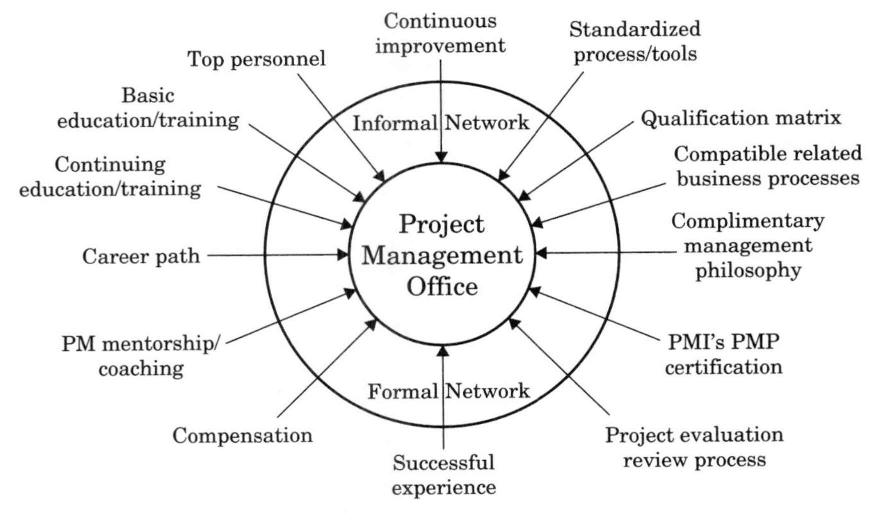

Figure 4.1 Environment for Project Success.

PROJECT PLAN ACCELERATOR

Using information from this chapter and the *PMBOK® Guide* for additional reference, make a note of any specific planning items that should become components of your project plan. Describe the component and develop questions around the component that will enable you to effectively use the component in the plan.

Project Plan Methodology—Development Steps

- Discuss the existing project management methodology in use by your organization with other project managers and potential stakeholders. Look for gaps and areas for improvement
- Schedule focus group sessions and needs analysis meetings to identify the specific concerns, requirements, and priorities of people and business entities that may be affected by the methodology.
- Gather information from other project-based organizations and identify lessons learned.
- Examine successful projects within your organization and identify the factors of success
- Create presentations to explain the benefits of a methodology.

Consider how the five process groups and the knowledge areas of the *PMBOK® Guide* can be used during the development of the methodology. Review and analyze the planning process in use by your organization. How well does the process work? Where can you make adjustments to improve overall efficiency? Look for models and procedures in use within your organization, including project dashboards, status reports, and planning procedures. In many cases you will find very useful, sometimes ingenious techniques created by a project manager or a project team for a specific project. These techniques could be adapted for use on the enterprise level. The items listed in the Project Plan Accelerator are provided as examples and to assist you with the development of actions that can be taken to achieve the desired results.

Notes:

Project Plan Accelerator

PMBOK® Guide Component / Tool / Technique / Process	*Action Plan for Implementation*	*Person Responsible*	*Date Required*
Project management methodology	Create a flowchart of the current methodology or identify common processes that appear to be repeated. Identify gaps, weaknesses, redundancies or other deficiencies. Simplify whenever possible.		
	Schedule focus groups and needs analysis sessions to identify concerns and priorities.		
	Obtain lessons learned from established project management offices.		
	Identify the benefits of an enterprise wide project methodology. Analyze existing tools available commercially. Explain what is attempting to be achieved through the introduction of a project methodology.		
	Create a change process and communicate the need for continuous improvement.		

Chapter Five

Defining Project Success

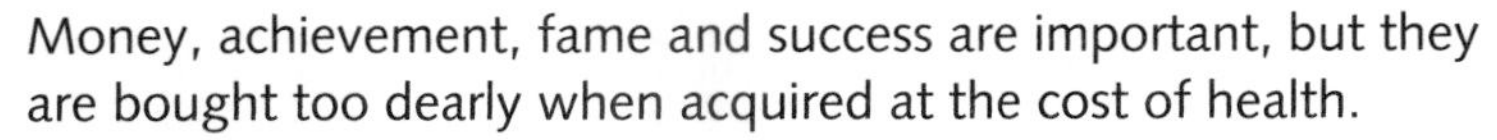

> Money, achievement, fame and success are important, but they are bought too dearly when acquired at the cost of health.
>
> —Anonymous

> Six essential qualities that are key to success: sincerity, personal integrity, humility, courtesy, wisdom, charity.
>
> —Dr. William Menninger

If you ask project manager to define the criteria for project success, the response would be, in most cases, "on time, on budget, and within scope or performance specifications. These criteria are often referred to as the **triple constraint**, shown in Figure 5.1. It is generally depicted as a triangle with cost, schedule, and scope representing each side. The Triple Constraint is intended to show the continuing challenge of managing these often competing project demands. Changes to any side of the triangle will impact one or both of the other sides. A quick example would be an increase in project scope. If the scope is increased, the cost and schedule sides should also increase proportionally (in a perfect world of course). But exactly how important are these three factors?

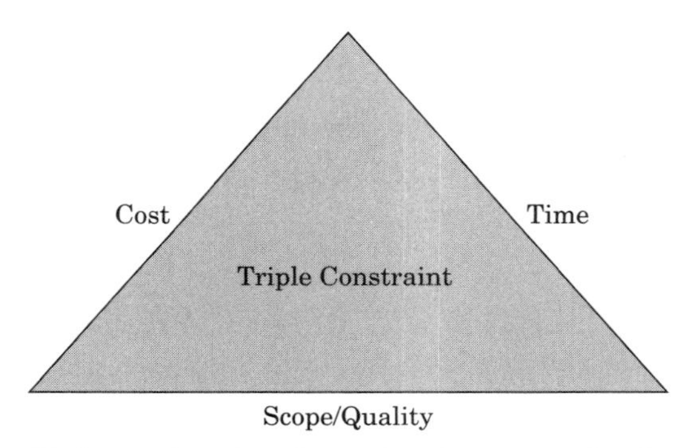

Figure 5.1 Triple Constraint.

Certainly these items need attention regardless of project size and complexity, and should be considered as part of the larger, more comprehensive view of success. It is important, however, for project managers and teams, and especially the sponsors and senior executives supporting the project, to identify the many other factors that define success. An important item to consider from every stakeholder's perspective—especially sponsors and project executives—is that projects are financial investments that should produce value to the organization and to the intended recipients. This value should be experienced immediately upon project completion and should continue to be experienced for a considerable time after the project has been closed. Projects should also be recognized as key elements when planning the strategies for achieving business excellence. This understanding of the financial implications, the possible benefits a project may provide, and the connection between projects and the organization's bottom line will help to define project management as an indispensable and critically important factor for success.

TYPICAL SUCCESS FACTORS FOR PROJECTS

Success factors vary by organization and may be defined in terms of **key performance indicators (KPIs)** or **critical success factors (CSFs).** There are probably at least a dozen acronyms used to describe success factors. Here is a list of success factors that should be considered when initiating a project:

- A strong and well-enforced planning process and a plan
- Clearly defined expectations and project scope
- A motivated team
- The willingness to say no (to unneeded changes, unreasonable requirements, unrealistic expectations)
- A well-defined and well-managed change-control process (to avoid scope creep and unauthorized changes)
- A risk management process that is supported by the entire project team and is engaged during the entire project life cycle
- A specific method for project closure to ensure that all project deliverables have been provided and accepted

The Standish Group, based in West Yarmouth, Massachusetts, is a research firm that focuses on mission-critical project management applications. It includes a definition of success in its studies about project management:

> The successful project is a project that is completed on time and on budget, with all features and functions originally specified. (This definition is common among project managers but we can define success in many other ways.)

The Standish Group also defines a challenged project and a failed project:

> Challenged: The project is completed and operational, but over budget, late, and with fewer features and functions than initially specified.
> Failed: The project is canceled before completion, or never implemented.

Considering all of the potential reasons a project can fail is a good idea. It focuses the attention of the project team on risk management and creates a sense of urgency. However, it is extremely important to maintain a concentrated effort on the many factors that are associated with success. This will help to keep team motivated to achieve objectives and will generally keep morale high. (A winning attitude is essential.)

Besides studying project failures, the Standish Group has also developed a recipe for success. Examining this list, you will notice that it includes many of the items described in Chapter Two, "The Big Picture":

- Executive support—this means visible and sustained support throughout the project
- User involvement—obtaining input and creating buy-in
- An experienced project manager—to create confidence within the team and with the other stakeholders
- Clearly defined business objectives—connecting projects to business objectives is critical for success
- Defined scope (minimized to the essential items)—Know what you are tasked to deliver
- A standard software infrastructure—the infrastructure should be well known by the project team and used consistently
- Firm basic requirements—a solid and well communicated requirements management process will minimize

issues and improve the probability of success delivery of project objectives

- A formal methodology
- Reliable estimates—Estimates developed by the actual performers of the work activities are generally the most reliable*

In the book *The World Class Project Manager,* by Robert Wysocki and James P. Lewis, the authors created a project manager profile to assist project managers in further developing their project management skills and competency. A strong set of project management skills can also be included in the list of success factors for a project. These skills include but are not limited to the following:

- *Fundamental project management.* Sizing and scoping of a project, scope statement development, understanding the process for activity duration and cost estimating, critical path development and management, developing and managing change control processes, monitoring and controlling tools and techniques
- *Personal skills.* Creativity, decision making, problem solving, team building and influencing skills, negotiating skills
- *Interpersonal skills.* Conflict management, balancing stakeholder needs, collaboration, communications skills—listening, writing, presenting
- *Business skills.* Developing budgets, assessing business needs and performance, understanding business processes, communicating between multiple business entities

---

*Source: Jim Johnson, Karen D. Boucher, Kyle Connors, and James Robinson, "Collaborating on Project Success," www.softwaremag.com/archive/2001feb/collaborativemgt.html.

- *Management skills*. Delegation, managing change, managing multiple priorities, organizing, time management, performance feedback
- *Testing*. Verifying and validating system performance, quality assurance

If you compare this list to the "Organization Process Assets and Enterprise Environmental Factors" described in the Project Management Institute's *PMBOK® Guide—3rd and 4th Edition,* you will notice a very close relationship to the critical success factors discussed in this chapter. Enterprise environmental factors include:

- Organizational or company culture—understanding the unique culture or an organization at the enterprise level and at the business unit level is a major factor in developing methodologies and plans for communicating information.
- Government or industry standards—an awareness of standards is essential in any planning process.
- Human resources—the skills, disciplines, and knowledge of the organization's people must be considered during project selection activities, planning, and determining outsourcing need.
- Organizational risk tolerances—an understanding of the organizations policies regarding risk management is another critical area for consideration when developing a methodology and determining plans for success. The risk adverse organization will define success differently from an organization that encourages significant risk taking.

Very fundamentally speaking, project managers are considered to be "general managers" (meaning they have a broad knowledge of many business processes besides project management) and therefore should possess a wide range of managerial skills that will ensure project success.

It is absolutely essential for project managers to understand that success is no longer limited to the Triple Constraint or "On time, within budget, and according to scope." Client satisfaction, profitability, impact on strategic objectives, reduction in costs, relationship to other projects and company operations, increased market share, and opportunities for new products are all part of project success. It is worth mentioning that project success may also be measured by a few additional items that are not always discussed during project reviews and closure meetings (but probably should be). While discussing a project with a group of project managers who recently completed a project and proclaimed it to be successful, I asked them to describe project success from their perspective. The observations were interesting and provided additional proof that success is measured in many ways and explained sometimes very creatively. One team's description of project success:

- The product or service is actually being used by the customer (a strong indication of success!).
- The project team survived (meaning they are all still employed).
- The project didn't make the papers (in other words: sometimes no news is good news). This particular project was a government project, and there was much sensitivity about failed projects finding a place on the front page of the local newspaper.
- The team is willing to work together on a new project. That is, the team worked together as a cohesive unit and managed to solve problems effectively, and they supported each other through the project life cycle.

- The team members are willing to work with the project manager again. This means the project manager demonstrated strong leadership skills and managerial skills and was able to gain the trust and respect of the project team.

In additional to these success factors we can add:

- Add on business as a result of the project
- New clients through references and reputation for success
- Clients will to provide their names willingly as a reference
- The new lessons learned and inspiring stories we hear about our projects. The wealth of new knowledge shared within an organization
- The willingness of team members to support each other
- The focus of a team working together to find solutions instead of looking for who is to blame

Today's measures of project success are significantly different than just a few years ago and the responsibilities of the project manager and the skills needed to achieve success have also changed considerably. Of all the success criteria and factors for success that have been discussed in this chapter, the greatest factor, in my opinion, is to truly listen to the client. When I was assigned my first large complex project, I followed the process developed by my company. It was a great process, and made a lot of sense. Afterall it was written by experts. I held the prescribed kickoff meeting, I defined the project deliverables, and I created a project binder and included all of the essential items that should be in a project plan. I met with the customer and proudly proclaimed, "Here's your plan!" The client gazed at me with a puzzled look and said, "Let me review this and I'll get back to you." The next day, the client said, "I can't use this, it's not what I need so now let me tell you what I want." I was shocked. I had followed all of the processes to the letter. How could the customer find a problem with it? I believed that I followed the process for guaranteed success. Or had I? The lesson learned here is that the greatest and most important factor that will increase your chances for project success is to ask the customer what success means to his or her organization. Failure to do this could certainly nullify any predefined definition of success you may have adopted and create an adversarial relationship with your customer. As Stephen Covey said "Seek first to understand, then to be understood." Excellent advice.

PROJECT PLAN ACCELERATOR

Using information from this chapter and the *PMBOK® Guide* for additional reference, make a note of any specific planning items that should become components of your project plan. Describe the component and develop questions around the component that will enable you to effectively use the component in the plan.

Defining Project Success

- Obtain or develop a list of success factors common to the projects your organization is involved with. Obtain the definition of success from a variety of stakeholders.
- Prioritize these factors from your perspective and from your team's perspective.
- Ask your client how success is defined and then validate the success factors you gathered from your organization from the client's viewpoint.

- Update the success factors you have developed to include the client's input and then communicate the new set of success factors with your team.

Consider how the five process groups and the knowledge areas can be used during the development of success factors for your projects. Make sure you obtain input from all key stakeholders to ensure you have a clearly defined view of success from a 360-degree perspective (the client, yourself, the team, the sponsor, and other stakeholders).

Notes:____

PMBOK® Guide Component / Tool / Technique / Process	*Action Plan for Implementation*	*Person Responsible*	*Date Required*
Project success factors Organizational of enterprise view	List the factors of success from each key stakeholder's perspective. Go beyond *on time, within budget, and according to specifications* to identify what success really is.		
Project success factors: The client's view	Meet with the client. Develop questions that will help uncover all of the elements for success. Attempt to identify hidden agendas and unstated expectations.		
Project success factors: The team's view	Schedule a meeting to discuss success factors. Use previous projects as examples. Uncover the hidden expectations and personal agendas of the team members.		
Project success factors: The sponsor's view	Obtain an executive or sponsor's view of project success. Connect the project to strategic objectives and ensure that all expectations have been communicated. Become aware of political issues that may be a factor in the planning process.		
The Triple Constraint	Communicate the basic elements of the triple constraint and how they affect project execution. Use this as a basis for defining total project success. Ensure that you discuss project success from several perspectives and go well beyond the Triple Constraint.		

Chapter Six

Developing Performance Measures

The process of developing performance measures is closely linked to the development and management of a project plan. It is important for the project manager and all key stakeholders to reach agreement about what will be measured and then how to actually measure performance. A vision and mission statement is developed at the organizational level, which provides a view of the bigger picture and a direction for the organization. Projects are formed and aligned to support and assist the organization in achieving the higher-level goals.

The *PMBOK® Guide* defines work performance information as follows:

- Schedule progress
- Project deliverables
- Activities in progress
- Completed activities
- Quality standards
- Project costs
- Estimates to complete activities
- Percent complete of activities

- Documentation of lessons learned
- Resource utilization

Other measures may include:

- Error rates
- Output per unit or through-put
- Compliance

Developing and communicating performance measures can be difficult for some organizations. There must be agreement about what will be measured, what is important, how to actually measure performance, and how to accurately and effectively communicate performance information. To facilitate the process, performance measures can be defined and placed into categories that align with the inputs, tools and techniques, and outputs of each project management process. Each element of the process can be measured by determining what inputs are needed and how much will be needed (when applicable, what was produced through the process), and how well the outputs were produced.

Examples:

- *Input measures.* Materials, equipment, staff time, and resources are inputs.
- *Output measures.* Products produced, services provided, (the deliverables) are the outputs to be measured.
- *Outcome measures.* The score associated with the outcome will reflect results in terms of accuracy or variance from actual target. Before the output is produced, the team involved in its development should have some specific criteria to refer to. This will minimize variances and keep the entire team focused on what should be delivered. Establishing outcome or output measures should be included in the planning process.

- *Efficiency measures*. These measures include the cost per unit produced, the number of outputs over time, number of outputs based on inputs. The team and the project manager should be aware of the effort required to produce the deliverables. This includes the resources, how the resources are used, what has been produced in terms of volume, use of available funds, unused material, or scrap.
- *Quality measures*. Reliability, safety, availability, maintainability, performance, fitness for use, and social acceptability (in some cases). Mapping out all key processes used in your project will assist in developing quality measures.

  After the measures have been defined, the next step is to control them by monitoring, measurement, and analysis. This will help to ensure that product quality objectives are met. If you can't monitor a process by measurement, the process should be reviewed to determine why measurements are not defined. Processes produce results, and results can be measured if a baseline has been established. The process itself can also be reviewed to determine if the appropriate steps are being followed and where adjustments are necessary.

The project management process has been defined as a series of processes that bring about a result. The results may be products or services or some type of deliverable that will be handed off to a customer or the next person in the process. (Refer to the customer supplier model in Chapter One.) Discuss the desired level of performance with the customer or stakeholder involved and determine levels of performance that are attainable and realistic. Providing examples of performance measures during the discussion about the deliverables produced during the process will facilitate the discussion and help to reach agreement about the final set of performance measures.

SETTING OBJECTIVES

Setting objectives is a critical element of the planning process. The project scope statement includes the specific objectives of the product and the project (see the project scope template in Chapter Twelve). Many organizations struggle to define clear, measurable objectives. Executives develop strategies and higher-level goals, which are then communicated downward to the various business units and departments. These goals are defined in increasing levels of detail as they descend through organizational levels, and eventually projects are created to support the higher-level goals. Specific, measurable project objectives are established to meet the higher-level goals. The actual development and communication of the detailed objectives often presents a problem for the project team, the client, or other stakeholders. The following information will provide some assistance in the development and refinement of project objectives.

Definition of Project Objectives

Project objectives are a series of specific accomplishments designed to address the stated problems or needs of the stakeholders and achieve the desired results or goals. An objective is an **endpoint,** the actual result, and not *a process*. It is a description of what will exist at the end of a project or phase of a project.

Ensuring the clarity and completeness of the objectives will enable the team to plan and implement activities that will lead to attainment of these objectives. Writing clear objectives also makes it easier to monitor progress and evaluate the success of projects.

When writing objectives, the project manager and team should avoid using *process* words. Process words are associated with an action that must be taken; instead, use *endpoint* words. Endpoint words focus specifically on what will be accomplished through a series of actions. Examples:

Process Words	**Endpoint Words**
assist	train
improve	distribute
strengthen	increase
promote	reduce
coordinate	organize
define	establish
assess	return
compare	design
analyze	
describe	
explain	

Objectives must be *specific* (what will be provided and when) *measurable* (how much, how many), and must describe what is *desirable* (what will be suitable and appropriate for the client, the organization, the situation) and *obtainable* (realistic; it can actually be accomplished with the resources and capabilities available).

The acronym **SMART** sums up the key components of a well-written objective. Table 6.1 describes SMART objectives.

Goals

A project goal describes at a high level the intended state to be achieved. This is basically what the environment or desired state will be at the project completion:

- *A goal is the solution to the problems you described during analysis or in high-level planning sessions.* The problem

Table 6.1 SMART Objectives

Letter	*Meaning*	*Purpose*
S	Specific	Is the objective clear in terms of what, how, when, and where the situation will be changed?
M	Measurable	Are the targets measurable? For example, how much of an increase or reduction is desired? How many items should be produced, or how many people will be trained?
A	Action-oriented	Does the objective specifically state what actions are required to achieve the desired result? In some cases the A refers to "attainable." Is the objective something that can be reached by the performers?
R	Realistic	Are the desired results expressed in a way that the team will be motivated and believe that the required level of involvement will be obtained? Is the description accurate?
T	Time-bound	Does the objective reflect a time period in which it will be accomplished (e.g., end of the first quarter or by end of year)?

statement is generally limited to those specific problems that could be solved by the project. Your goal statement represents the solution.

- *A goal is realistic.* It is not intended to state more than the project can possibly achieve.

It is important to develop a goal statement before developing and defining the scope of the project.

Process for Developing Performance Measures

This process can be used during discussions with a project team or stakeholders to define the performance measures for a project and to ensure that the correct items are being measured using the appropriate criteria. Be sure to review existing documentation about performance measures. These may be found in your Organizational Process Assets. Don't try to rewrite processes that are actually working very well. At some point, you may want to analyze these processes for potential improvement. You may also find that one department in an organization has a well-defined process that has not been communicated to other departments. The amount of information available within an organization that has been developed by the employees and has not been distributed can be surprising and can be extremely useful (and time saving). Project managers should routinely seek information from other areas in the organization or other project managers. Undocumented lessons learned are in abundance in most organizations. A common practice is to schedule "lunch and learn sessions," where project managers gather to share information about projects, specific experiences, and the occasional "war story."

Definition of Performance Measures

Performance measures are quantitative descriptions of the quality and capability of products and services offered by an organization.

The process of defining performance measures varies by organization, but there are a number of similarities. The following summarizes the typical steps used to define and document project or other organizational performance measures: This process was adapted from the Sandia National Laboratories approach. A six step process that provides a straightforward and logical process for defining performance measures. The approach was used as a model to create a process for developing project performance measures. Performance measures in the *PMBOK® Guide 3rd and 4th Edition* include such items as schedule and cost data, deliverable completion, percent physically complete of in progress activities

Step 1. *Describe the desired outcomes or the outputs*. In the project environment, there will most likely be several targeted outcomes or deliverables. The question associated with this step is: Why are we doing this work?

Step 2. *Describe the major processes involved*. Review organizational process assets and determine which of these processes must be included in the planning and execution of the project. Define clearly what will be done and how it should be done, including which organizations will be involved. Review the nine knowledge areas of the *PMBOK® Guide* and the associated processes. Which processes will be used during the project life cycle?

Step 3. *Identify the desired results*. What should be produced in terms of products, services, or other deliverables? These may also include the outputs of a specific process, if there is a need for very detailed planning. Extreme detail may be required for very-high-risk projects. The targeted stakeholders must be able to describe what the desired results are in order to ensure that the team focuses on the real needs and will be able to produce them or improve them.

Step 4. *Establish performance goals*. The team should be able to determine when the desired result has been achieved. A quote that comes to mind here is "projects progress quickly until they are 90 percent complete and then they stay 90 percent complete forever." The team should know when they have achieved their objectives. Without a clear understanding of what *completed* means, it will be difficult to actually reach the point of closure. Make sure you define the acceptance criteria for the project from the client's perspective and the team's perspective

Step 5. *Define the measures for the objectives that have been set*. What will be used to track and measure progress? Organizations that have established project management offices will have a set of standardized measures for projects. These measures will have certain characteristics that make them useful, discussed next.

Examples of Performance Measures

Depending on the project and the goal, various performance measurements can be used:

- Income compared with previous year / quarter
- Volume—Increase or decrease in output
- Funding—Comparison with budget requests and actual costs
- Cost savings—Percent of reduction in cost
- Spending (comparison with previous year)
- Revenue increase
- Customer satisfaction levels
- Quality levels or effectiveness of services or programs
- Physical facilities—Expansion to meet attendance and to improve services and networking or member– stakeholder interaction
- Number of publications produced
- Awards received
- Number of educational hours provided
- Number of scholarships granted
- Expansion of services
- Participation increases

- Number of initiatives in progress
- Number of projects cancelled
- Number of projects completed late

Step 6. *Determine the required metrics*. What specific inputs, processes, or outputs should be measured? Also consider the baselines that will be needed for comparison. Baselines provide the reference points that can be used to determine where variances are developing and whether or not the variance is within acceptable tolerance levels. Failure to establish baselines creates a situation where comparisons between actual performance and planned performance can not be accomplished resulting in a lack of useful data to gauge project performance or overall project health.

Characteristics of Performance Measures

Measures are selected to track progress and provide indicators about where changes must be implemented to improve a process or correct an undesirable variance. Measures will also assist in identifying processes that are not producing results at an acceptable rate and determining actions that can be taken to improve the rate of performance. These are the characteristics of performance measures:

- They reflect results, not the activities used to produce the results.
- They relate directly to a performance goal.
- They are based on measurable data.
- They use normalized metrics for benchmarking.
- They are easy to understand.
- They create an opportunity for continuous improvement.
- They are cost effective.
- They are accepted by the stakeholders and have owners.

Performance Metrics

Many organization use key performance metrics as tools to determine whether performance measures are being met:

- Stakeholder confidence
- Stakeholder expectations met
- Board of directors' acceptance of short- and long-term directions
- Strategic objectives and goals attained/fulfilled
- Strategic objectives' progress toward goals
- Organizational goals and targets defined
- Employee satisfaction/motivation
- Shareholder perception of strategy effectiveness
- SWOT analysis
- Residual risks determined from audits
- Comparisons of internal strategy development process to leading external strategy development processes
- Planning process objectives supported by benchmarks and comparisons
- Supplier and partner capabilities
- Information technology capability to support strategic objectives
- Conformance audits
- Accumulated performance measures related to the strategic action plans

Examples of Metrics

- Number of programs this year compared with the number of programs last year
- Funding available compared to funding requested
- Customer satisfaction this year and customer satisfaction last year

- Number of projects completed on time compared with the number scheduled
- Total actual score compared against the total possible score (this could be used for measuring quality, customer satisfaction, or other evaluations)
- Amount spent to date compared with amount budgeted (earned value analysis)*

PROJECT PLAN ACCELERATOR

Using information from this chapter and the *PMBOK® Guide* for additional reference, make a note of any specific planning items that should become components of your project plan. Describe the component and develop questions around the component that will enable you to effectively use the component in the plan.

Developing Performance Measures

- Research the performance measures that currently exist in your organization. How is project success determined? What tools are used to measure actual performance? How well are processes for measuring performance documented and understood? Where is performance information stored or archived?
- Develop a check list for determining the "health of your project." How do you determine how well your project and the project team are performing? How do you know when a project is beginning to show signs of problems? What are the warning signs?

---

* Source: "Developing Performance Measures—A Systematic Approach," www.assess.sdes.ucf.edu and www.orau.gov

- Determine what performance indicators and indexes are in use. What additional indicators are needed? If you were to develop a project performance dashboard what gauges or indicators would you include?
- Identify the customers (the receivers of the products, services, or deliverables). Customers are a great source for defining success factors and performance expectations.
- Decide on the outcomes desired, and discuss how these outcomes will be achieved. Make sure you consider the desired outcome and the planned outcome. Sometimes these can be quite different.
- Place emphasis on the people involved. They should have a voice in the development of performance measures and should be included in the formal approval and acceptance of these measures.

Consider how the five process groups and the knowledge areas can be used during the development of performance measures. Review existing documentation. Determine what really should be measured.

Notes:____

____

____

____

____

____

____

____

____

____

____

____

PMBOK® Guide Component / Tool / Technique / Process	*Action Plan for Implementation*	*Person Responsible*	*Date Required*
Work performance Information	Determine who will be involved in the measurement process. Examples of work performance information: • Schedule progress • Completed deliverables • Deliverables not completed • Milestones achieved or missed • Quality—number of defects, service level fluctuations • Costs planned and incurred • Percent complete • Comparison of estimates against actual results		
	What are the items that should be measured for the project? What are the expectations of the sponsor or project executive? What information or lessons-learned documents are available? Schedule interviews with key stakeholders to identify performance measures.		

<table>
<tr><td></td><td>How will the outcomes be measured and who will measure them? When will performance be measured and how will the results be communicated?</td><td></td><td></td></tr>
<tr><td></td><td>What are the criteria for success?

• From the team's perspective
• The client's perspective
• The sponsor's perspective
• The End user's perspective</td><td></td><td></td></tr>
</table>

Chapter Seven

Monitoring and Control Simplified

Project monitoring and control is an essential element of the project management process. Without control, there would be confusion, scope creep, scope drift, and scope "leap" and a considerable amount of conflict among project stakeholders. Chaos would become the norm throughout the project life cycle. Measuring and monitoring project progress using a specific process increases the ability to identify symptoms of problems and trends toward unacceptable variances. Failure to do this on a regular basis could result in project delays, poor quality, safety risks, and increased costs just to name a few. The monitoring and control process is part of the overall project planning process and begins with the development of mutually agreed upon performance measures and procedures for tracking and managing the project. These mutual agreements between the functional entities involved and other key stakeholders should be established early in the project life cycle and reviewed and updated on a regular basis. The closing of a project phase is an ideal time to review the project's current condition and update the existing monitoring and control procedures. The lessons learned from each phase can help to prevent the re-occurrence of issues and problems and prevent unnecessary conflict and stress. During each phase, the project team should be continually assessing results and making comparisons with the original plan looking for symptoms and potential problems. Of course the project manager should also be looking the positive indications as well. Project managers should plan to recognize their teams when variances are minimized and projects appear to be in control. This is a critically important factor. Project managers will, regardless of how well planned a project may be, experience unexpected situations. The monitoring and control procedures will assist in developing appropriate responses. The team and the project manager will also make daily decisions about work activities and how resources will be used to overcome potential threats. The monitoring and control process, when properly managed will generate some very important lessons learned for future use.

Figure 7.1 illustrates in a very simple, straightforward manner how the monitoring and control process works. It may seem like an oversimplification of the process, but a close look will reveal that it shows very clearly what typically occurs as a project is monitored and controlled.

The first step in the process is to have an agreed upon, approved, and supported project plan. The plan is the result of a collaborative effort between the key stakeholders with emphasis on the desired deliverable as well as the impact to other stakeholders not directly involved in the project. When the project plan is approved and the go ahead is given, the project plan is executed and the process of monitoring and control begins. The process is actually cyclical in nature. It repeats itself throughout the project life cycle.

Execute the plan. The resources are engaged, contractors are selected and contracts are signed, Goods and services

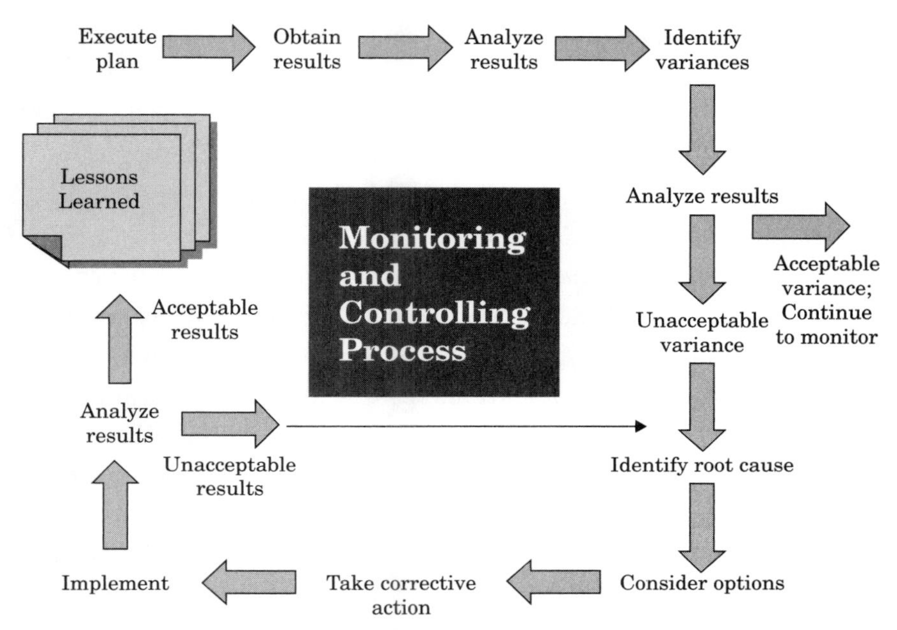

Figure 7.1 Monitoring and Controlling Flow Chart.

are procured and the work force (performers) begin their assigned activities.

Obtain results. The execution of the plan will, after a short period, begin to produce results. As work is completed, data are provided for analysis.

Analyze results. Compare the results against the plan baselines. The objective is to determine if there are any variances to the plan.

Identify variances. During the analysis, the reviewer (the project manager) identifies any variances during the analysis process. Variances are deviations from the plan and could develop into very serious project situations.

Acceptable variance. The variance is within acceptable levels. Monitoring and controlling should continue.

Unacceptable variance. The variance is outside acceptable levels. and additional action is required. Failure to act on this variance could jeopardize the project in several areas. Project results are generally integrated. and interdependent and problems in one area could have an effect at the project level.

Identify root cause. A common approach when a problem or variance is detected is to immediately apply a solution. This "jump to solutions" approach can lead to inappropriate or less than optimal actions to solve a problem. Taking the time to identify the root causes of a variance will generally lead to better decisions regarding how the problem will be solved. Use of a **fishbone** or **cause and effect diagram** is extremely helpful in narrowing the key root causes.

Consider options. Before deciding on a response, the project manager and project team should identify and discuss options. Encourage some innovation and creativity and think about the alternatives. There is usually more than one way to solve a problem. Consider the risks, the cost, the type of resources that may be needed, the time required, and the effect the solution may have on the project as a whole and on the organizations involved. Be sure to keep the stakeholders in mind during the decision process. Remember, the best solution may not be the most obvious. Use whatever available time you have to generate solutions. Even during crisis situations it is a good practice to look at options first before making decisions.

Determine the corrective action. The process of determining options will result in a choice of actions that may be taken to resolve the variance. The team selects the option that has the greatest chance of resolving the problem, taking into consideration the risks and costs associated with the decision. Careful review of the options will result in the selection of the best approach to resolve the situation. A defined set of measurements should also be developed in this process.

Implement corrective action. The selected option is executed through a specific agreed upon plan. How will it be executed, who will do it, and who is responsible for completion of the actions are questions that should be reviewed.

Analyze results. During and after execution of the approved corrective action, the results of the action are reviewed to determine if the desired condition has been achieved. Clearly defined metrics and measures of success should be in place and utilized when analyzing the results.

Unacceptable results. Occasionally, the outcome of a corrective action will not be acceptable. The variance may not be reduced to satisfactory levels and additional action is required. This may require a return to root-cause analysis to ensure that the true root cause has been identified.

Acceptable results. If the analysis of the corrective action indicates that the problem has been solved, the variance has been reduced to acceptable levels, or the issue has been eliminated, the corrective action process for the situation is closed.

Lessons learned. It is generally considered a best practice to review the findings of a corrective action and document what has been learned for use on future projects or if the situation should occur again during project execution.

Monitoring and controlling is a critical process in the project life cycle. It involves ***Integrated Change Control*** the larger more comprehensive view of change management and configuration management that includes the relationships and impact of change between all project management knowledge areas. Integrated change control assures the team is aware of how a change in one area may have an impact on several, if not all of the major knowledge areas of project management. The subsets of Integrated change control include: scope control, scope verification (the acceptance of the work performed), schedule control, cost control, quality control, managing the performance of the team, reporting status, managing risk and risk responses, and contract administration. Developing a simplified approach to monitoring project performance or any other process associated with managing a project should be part of the overall planning process. *Simplified* does not mean "simply" encouraging risky short cuts. It means analyzing a process and working with a team to determine the best approach with the most efficient steps that will produce the desired outcome without adding burdensome administrative work.

Mature project management is associated with the development of processes that make sense for all stakeholders while minimizing the administrative burden that sometimes accompanies a formal project management process. The use of flowcharts, systems mapping, and other techniques reduce the need for wordy, complicated documents that are often ignored by the project team and stakeholders. The common-sense approach to project management processes is to discuss what is needed, find an effective way to communicate the need, assign people to do it, give them the tools they need, define acceptance criteria for any deliverables that will be produced, and then enable the resources to perform. Putting in place clearly defined roles, a good plan, and agreed-upon processes will help to reduce the effort associated with monitoring a control and increase the probability of successfully completing the project.

PROJECT PLAN ACCELERATOR

Using information from this chapter and the *PMBOK® Guide* for additional reference, make a note of any specific planning items that should become components of your project plan. Describe the component and develop questions around the component that will enable you to effectively use the component in the plan.

Monitoring and Controlling

The monitoring and control process should be developed in steps. An ideal time to develop these procedures is during the project kick–off meeting. This meeting provides the key stakeholders with an opportunity to obtain detailed information about the project, learn about the roles and responsibilities of the other team members, and reach agreement about how changes to the project plans will be managed. The following are possible steps that may be taken in the development of agreed upon monitoring and control procedures.

- Identify existing control procedures. Review the organizational process assets currently in place. The *PMBOK® Guide 3rd and 4th Edition* provide a very useful list of typical organizational process assets.
- Determine what tools will be used for monitoring and controlling the project:
  - Project software
  - Earned value analysis
  - Control charts
  - Pareto diagrams
  - Cause and effect analysis

    - Product sampling
    - Project reviews
    - Team reviews
    - Status reports
    - Audits
    - Walk-throughs
- Determine stakeholder needs. Stakeholder requirements regarding status will vary, and it will be necessary to identify the priorities of the key stakeholders and their specific expectations associated with monitoring and controlling. Conduct a stakeholder assessment to ensure that all stakeholders have been identified. Determine the influence level of each stakeholder and their interest in the project.

Consider how the five process groups and the knowledge areas can be used to develop and then manage the monitoring and controlling process. Identify what will be monitored, determine acceptable variance levels, and decide what tools will be used for monitoring and controlling. Establish an escalation process that is acceptable to all stakeholders, and make sure the process is followed.

Notes:____

Project Plan Accelerator

PMBOK® Guide Component / Tool / Technique / Process	*Action Plan for Implementation*	*Person Responsible*	*Date Required*
Monitoring and controlling	Identify the specific control needs for the project. Consider the expectations of the sponsor and the client. Develop a control process for the project based on organizational needs, project specific needs, and the processes currently in place. Use the project team or select individuals from each entity involved to assure buy-in to the process. Communicate the process for monitoring and controlling the project. Obtain feedback on the process and revise as necessary to obtain stakeholder approval and ensure compliance. Analyze current procedures and determine where improvement is necessary. Review past project performance to identify critical monitoring and control areas. Review current organizational process assets.		

(continued)

Project Plan Accelerator (*Continued*)

PMBOK® Guide Component / Tool / Technique / Process	*Action Plan for Implementation*	*Person Responsible*	*Date Required*
Project management plan Escalation plan	The project management plan documents the set of outputs form the planning process and provides guidance about how the project will be executed, monitored and controlled. Determine how well the project management plan is defined. Identify gaps and potential weaknesses and develop additional input. (Caution—don't fall into the trap of a spiraling planning process. Focus on the project needs and develop the plan accordingly. Develop a plan for escalating issues that cannot be resolved at the project level. Ensure stakeholder agreement and commitment to this process.		
Performance Measurements	Determine what must be measured, define acceptable variance levels, and select the tools and techniques appropriate for the type of project.		

Integrated change control	Review the existing control process. Ensure that an integrated, systems type approach for managing change is in place. Develop any additional items to meet the needs of the project. Determine how the change control process will be communicated.		
Scope Verification	Determine how scope verification will be implemented. Scope verification is associated with reviewing and accepting the work and results that have been produced. Develop a process for comparing results with expectations and the defined and approved project scope of work. Ensure that the team understands the need for scope verification. Ensure that a process for formally accepting all project the deliverables has been developed and communicated.		

Chapter Eight

The Change Control Process

Stakeholders, especially clients, often change their mind about requirements, project scope, and just about anything that is planned to be delivered. Changes in the business environment, such as mergers, changes in strategies, cost cutting programs, and reorganizations after a project has been initiated will generally have a significant impact on the initial set of planning assumptions and can be very traumatic to the project and the project team in cases of mergers/acquisitions or changes in the executive management team. These types of changes often require the project manager and team to reexamine the scope of work and original agreed-upon deliverables and then work with other stakeholders to reach agreement on a new set of deliverables and assumptions. A process is needed to ensure that only those changes that are necessary and beneficial (sometimes regulatory changes do not translate into benefits) are introduced into the project plan. Generally speaking, if all changes were accepted, the project would quickly overrun its budget and schedule and might never actually come to an end.

Consider this:

"If change is allowed to flow freely, the rate of change will exceed the rate of progress". (*source unknown*)

This quote pretty much sums up the need for a **change control process.** Without a process to manage change, there would be no appreciable progress and basically an unending project. Control of the project simply would not exist.

The *PMBOK® Guide* introduces change control in several ways: integrated change control, configuration management, scope change control, schedule change control, quality control, risk monitoring and control, and contract change control. Basically these change control elements are subsets of Integrated Change Control and are often referred to as subsidiary plans to the project management plan.

All change control processes are associated with the need for an **integrated process for managing change.** Each change control process within a project is linked, and the project manager and project team should be aware that changes in one area, as an example a change in the scope of work, may impact several other areas, such as schedule and cost. This is a direct relationship to the concept of the *triple constraint*. Project management is an integration of systems in the form of the nine knowledge areas and five process groups. The linkages between the knowledge areas must be monitored to maintain awareness of the impact of decisions made as the project progresses. There is a relationship between change control and risk management in that the consequences of a decision should be understood before the decision is implemented.

The very core of a change control process is the impact analysis component. It is closely related to project risk management. In risk management, the project team assesses the probability that a risk event will occur and impact of that event if it actually occurs. Similarly, impact analysis is intended to carefully assess how the change may affect the project implementation process, the deliverables of the project, the organization sponsoring the project, other projects, and the general operations of an organization. When engaging the change control process, the project team assesses the impact of the change across all levels, phases, and major project components. This is important to provide assurance that the change will produce the desired effect and minimize any negative results.

There are eight basic elements of a change control process:

1. Establishing and using baselines to determine if changes have occurred. Examples would be schedule slippages and budget overruns or scope unauthorized scope increase.
2. Understanding the factors that may introduce change—these factors include changes in strategic direction, funding constraints, economic uncertainty.
3. Identifying the potential sources of change—change can originate from most project stakeholders and also from external factors such as government or changes in industry standards.
4. Reviewing and approving submitted changes—this is where impact analysis is engaged. All change requests are analyzed for value as well as need. Some changes will be required to satisfy compliance issues and regulations.
5. Managing approved changes—approval of the changes by some type of steering committee or change control board.
6. Updating the project plan after a change has been approved. The project management plan will change several times as the project progresses through each phase. This does not necessarily mean that the scope of

the project will change, but the method to achieve the result may change many times.

7. Documenting the lessons learned associated with the change—an often forgotten step in the change management process. Documenting lessons learned can save the project team and the organization a significant amount of time and funding when planning future projects.
8. Rejecting changes that are not considered beneficial, cost-effective, or that have other negative impacts. This is another area that is not generally well managed. A record of rejected changes is necessary to prevent them from being resubmitted. The rationale behind the rejection should be documented, communicated, and retained for future reference.

Change control also includes configuration management. This is generally associated with the features, functions, and physical characteristics of a product. Changes to a product's configuration should follow a specific process to ensure that changes are communicated and to prevent damage to equipment and unsafe conditions for people performing the work. Changes in the actual configuration of a product will typically impact a number of other areas such as scope, time, quality and budget. A careful assessment of the configuration change can save a substantial amount of unneeded additional work or rework.

The integrated change control system as described in the *PMBOK® Guide 3rd and 4th Edition* is a collection of formal documented procedures that define how project deliverables are controlled, changed, and approved. To be more specific, it is a system that provides a process for stakeholders to introduce a change, have a team analyze the change for benefits and approve or deny the change, track the change through implementation, and then analyze the results of the change.

By establishing a change control process, the project manager and project team (or the change control board, if one has been established) can make decisions about whether a change should be approved and implemented. If the change is rejected, the process explains how to communicate *why* the change was not accepted. The system provides the means to document approved or rejected changes for future reference.

The change control process, if communicated effectively and *practiced,* increases the probability of project success by minimizing (hopefully eliminating) unnecessary changes, providing a means of analyzing the rationale behind a change, and then communicating that information to the stakeholders.

MANAGING PROJECT CHANGES

For the best chance of buy-in from the project team, the change control process should be introduced during the early stages of the project, preferably at the kickoff meeting. The kick-off meeting, by my definition, is the formal beginning of the project planning process and where the key stakeholders and project functional managers or technical experts meet to discuss issues and collaborate about the most effective methods and processes for achieving project objectives. Many organizations do not have a formal and documented change control process. In keeping with the goal of providing simplified techniques, the following is a list of items that can be considered when developing a change control process or modifying an existing process:

- Assess the project team's change readiness. How does the team respond to changes? You are likely to find some resisters within the team or among the population of project stakeholders. Attempt to uncover the main reasons for their resistance before forcing changes upon them.

- Develop a change management strategy for the project. Review organizational change processes and modify as needed to meet the need of the project and its stakeholders.
- Identify and train (as needed) the change control team or change control board—the training may be associated with decision making skills, analysis techniques, effectively communicating ideas for change, updating people about organizational strategies, developing a "big picture" view when assessing changes.
- Provide an overview or training session for the sponsor or managers—make sure the sponsors and executives associated with the project are aware of the change process and its rationale.
- Document the change process (use flowcharts and diagrams whenever possible). Visual examples are often easier to explain and to comprehend.
- Plan to audit the process to ensure compliance—regular reviews to encourage process improvement are always a good idea.

The change control process is basically a step-by-step flow of information and actions (see Figure 8.1). The change control team receives a change, analyzes it, determines the risks and the benefits, makes a decision, and if approved, monitors the change for the desired result. Figure 8.1 shows the process that may be followed to document changes and ensure that the appropriate or required changes are introduced into the project. The following is a brief review of each step:

Step 1. *Identify the source of the change*. Changes may be initiated by the client, the government, the project team, the team, sponsors, functional groups, and other entities.

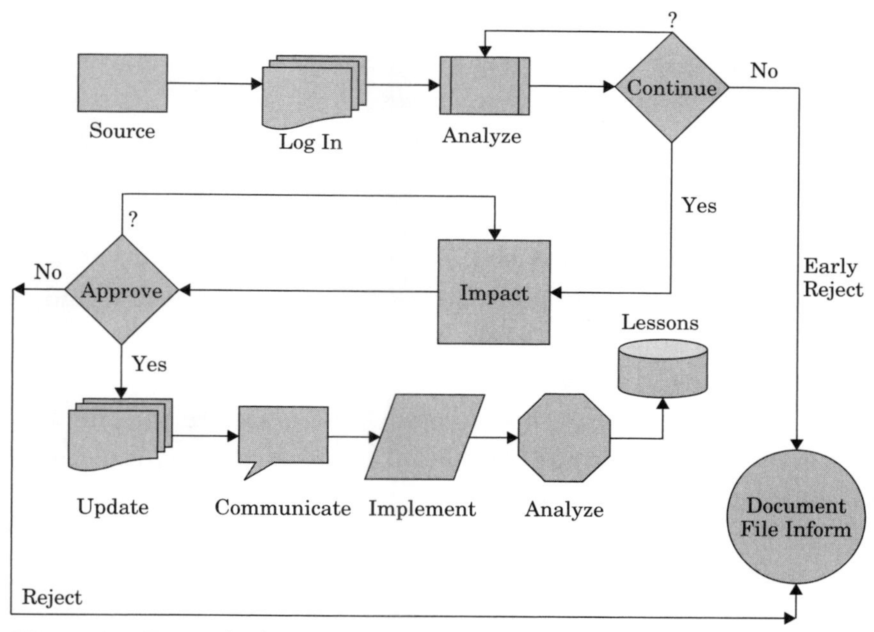

Figure 8.1 Typical Change Control Process.

It's a good idea to become familiar with each source of change and their specific biases and priorities.

Step 2. *Log in the change*. All changes requests should be documented, using some type of official change request form. Verbal change requests should be discouraged if not completely prohibited. In this step, the requestor of the change provides a form or some type of official document that includes the name, contact information, description of the change, and rationale behind the change.

Step 3. *Analyze the change request*. This is actually an optional step in the process. For some projects, especially large, complex projects, the stakeholders will generate several changes throughout the project life cycle. Some changes will be trivial and others will involve work that is outside the scope of the project. During this analysis, the change request is reviewed from a high level. The focus is on who will be involved, the benefits of the change, the risks, and the cost of the change. It is basically a screening of the change requests to determine which changes should be considered for more detailed analysis. The yes decision indicates that the change should go forward for a more detailed analysis. The no decision indicates that the change is too risky, too costly, or not necessary, and resources and time should not be allocated to the change. The change request received an early reject. The reason for the rejection is documented, the source is informed, and the request and the details around the request are filed for future reference. The question mark indicates that more information is needed and is sent back for more analysis.

Step 4. *Determine if the change should continue or decide to deny the change before any detailed analysis is conducted.* Some changes will be determined to offer very little in terms of benefits and may be very risky or costly. This early deny of a change should be documented for future reference. If the change control team cannot decide, they would go back into the process for further analysis. If the change control team decides that the change has potential benefit, the change request follows the next step in the process—Impact analysis.

Step 5. *Conduct impact analysis*. This step is the heart of the change control process. The change request is analyzed in detail to determine the following:

- Risks associated with the change
- Cost of the change
- Impact on quality if implemented
- Effect on customer satisfaction
- Impact to the organization if the change is implemented

- Impact on team morale
- Effect on other projects
- Effect on organizational operations
- Impact on schedule

When the analysis is completed, another decision step is required. This decision step involves the following:

A "No decision". This indicates that the change will not benefit the project or has too many negative possibilities. It is denied. The reasons are documented, the source of the change is informed, and the information is filed for future reference.

Question mark. The change control team cannot make a decision and additional analysis is required. The process goes back to impact analysis for further discussion.

Yes decision. This indicates that the change request should be approved and the change control team feels that the change is necessary or beneficial and has the supporting detail behind the decision.

Step 6. *Update the plan*. The approval of the change requires an update of the plan itself. The documents associated with the approved change are revised to include the change decision and other supporting information.

Step 7. *Communicate the change*. The decision about the change does not always include input from the project team or other affected stakeholders. Changes are often resisted and require explanation and an attempt to gain buy-in from the stakeholders who will have to implement the change or will be affected by the change.

Step 8. *Implement the change*. The change is implemented in a coordinated manner to ensure minimal disruptions to other activities. The change control process should include some guidelines about the actual implementation of the change. This is required to reduce the probability of introducing a problem into the project.

Step 9. *Monitor the results of the change*. As it is implemented, the change is observed to determine the actual results. The focus is on what has been accomplished, what benefits have been gained, and whether the change succeeded in correcting a problem. If the desired results are not attained, the change control team will analyze the results and modify its strategy. A careful analysis of the change and the potential outcomes will minimize probability that the change will not be successful.

Step 10. *Lessons learned*. Anytime a change is implemented or an action is taken to correct a problem, the team should analyze the process and document what has been learned. This may seem like an optional, even trivial, part of the process, but the value of lessons learned has been proven time and again. Take the time to review what has been learned and share with your organization.

PROJECT PLAN ACCELERATOR

Using information from this chapter and the *PMBOK® Guide* for additional reference, make a note of any specific planning items that should become components of your project plan. Describe the component and develop questions around the component that will enable you to effectively use the component in the plan.

The Change Control Process

It is always a good idea to review what is currently available before initiating the development of a new process. The current process may be completely satisfactory or may only need some minor adjustments. Always work form a point of facts and reliable data.

- Obtain information from other projects or from organizational process assets about how change is managed. Is there a standard change control process in place? Determine if it will work for you project or if revisions are necessary. Remember, every project is unique so make sure the process fits the project.
- Develop a change control flowchart or modify an existing process and communicate the process to the team. Flow charts can be very effective when communicating how a process works. Change control is basically a common sense component of the project plan and a straightforward, uncluttered process will be much easy to "sell" to the team.
- Emphasize the importance of a change control process and revisit the process on a regular basis during project status meetings.

Consider how the five process groups and the knowledge areas can be used during the development and implementation of the change control process. Attempt to involve and gain support from the stakeholders involved in the project. Explain the change process in terms the team will readily understand and accept. As you progress through the project life cycle identify, on a regular basis, the condition of the project in relation to its baselines. Stakeholders should know where the project is, in terms of status, at any particular moment. If a change is introduced, determine what the change is intended to accomplish and how the results of the change will be measured. Finally, communicate information about the change to the people involved. Include stakeholders in discussions about change whenever possible, encourage feedback from the people involved, and share information openly. Take the mystery and the anxiety out of change control. The whole process is really just common sense

Notes:____

____

____

____

____

____

____

____

____

____

____

____

PMBOK® Guide Component / Tool / Technique / Process	*Action Plan for Implementation*	*Person Responsible*	*Date Required*
Scope control Schedule control Cost control Quality control Risk monitoring and control Contract Administration Communications Management (These are or could be considered subsidiary plans to the project management plan)	Review existing processes for controlling change. Identify available change documents and procedures. If there is a PMO within the organization, obtain any standard procedures for managing change. Obtain best practices and lessons learned from previous projects. Conduct a change management best practices meeting to uncover techniques that have been effective for a project team but have not been communicated.		
Integrated Change Control	Verify that baselines have been established for use in managing and detecting change.		

(continued)

Project Plan Accelerator (*Continued*)

PMBOK® Guide Component / Tool / Technique / Process	*Action Plan for Implementation*	*Person Responsible*	*Date Required*
Configuration Management	Review and document how changes are approved and what the process is for implementing approved changes.		
Contract Administration	Determine how to manage rejected changes to ensure that they are not implemented.		
Project Management Methodology	Define a change control process for the current project. * Recommendation—Avoid developing detailed, lengthy text documents if possible. Use charts and graphics to explain the process.		

Chapter Nine

Establishing Roles and Responsibilities

The *PMBOK® Guide* chapter about the knowledge area: "Human Resource Management" and Chapter Two: "Project Life Cycle and Organization" address the need for the project manager to understand the roles and influence levels of a project's key stakeholders, the importance of developing the project team, and the generally accepted roles and authority levels of the project manager. The Human Resources Chapter (Chapter 9 in the 3rd and 4th Edition of the *PMBOK® Guide* focuses attention to the staffing, development and management of the project team. There is great emphasis on roles and responsibilities and the connection to project success. Interviews with project managers indicate that one of the major reasons for project failure is the lack of clearly defined roles for each project team member and key stakeholder. A failure to assign roles and obtain agreement from all team members about assignments and responsibilities will lead to confusion, conflict, and in many cases, a total breakdown of the team. In addition to assigning roles, the project manager must develop an understanding of how team members interact with each other and with other stakeholders and to recognize how personality and communication styles impact the effectiveness of the team.

To manage the many types of personalities and to ensure successful project completion, the project manager must assume a number of different roles and could possibly wear several "hats" on any given day. Changes from role to role daily and during the various stages of the project is quite common for project managers. There are at least twenty generally accepted roles and responsibilities of the project manager:

1. Demonstrating accountability for overall project success—the project manager accepts accountability for the success or failure of the project. Accountability can not be delegated. The project manager must commit to the project.
2. Planning for, acquiring, developing, and managing the project team—this includes coaching, mentoring, teaching, and providing feedback.
3. Assuring compliance with all conditions of the project—knowledge of the contractual items, when applicable and other defined and documented expectations and criteria.
4. Scheduling and managing project meetings and ensuring preparation of agendas and meeting minutes.
5. Scheduling and chairing the project kickoff meeting—this is a critical step in the overall project management process and, when effectively managed, creates a solid planning foundation.
6. Facilitating the development of the following: scope statement, work breakdown structure, responsibility assignment matrix, master schedule, escalation process, and processes for monitoring and controlling the project.
7. Negotiating with and obtaining approval from the client regarding scope of work, schedule, and client responsibilities.

8. Establishing teams or committees to develop project management subsidiary plans when applicable:
   - Implementation plan
   - Quality assurance plan
   - Change management plan
   - Escalation plan
   - Project budget
   - Training plan
   - Testing and acceptance plan
   - Cutover plan (depending on type of project)
   - Client acceptance plan
   - Operations, administration, and maintenance plan (These are the specific written directions and support information that will be provided to the user of the product or service. They are the post project environment guidelines that ensure a smooth transition and continued performance of the product after acceptance.)
   - Safety plan
   - Disconnect plan (This is associated with projects that involve the replacement of equipment and systems.)
9. Implementing formal project management processes and utilizing tools and techniques to manage project performance:
   - Use of project management software and enterprise systems
   - Integration of the major components of the project including subprojects
   - Development and management of a time-phased budget
   - Development of contingency plans
10. Monitoring project activities through project team meetings, obtaining project status and current health of the project, and reporting project status to sponsors, clients, and key stakeholders.

11. Verifying adherence to project plans and successful completion of milestones.
12. Reviewing contract implications and maintaining awareness of contract risk items, including penalty clauses and other terms and conditions.
13. Promoting the identification of problems and issues, and promoting organized and effective methods to solve those problems and conflicts.
14. Escalating issues as necessary to support project team members and issues that require executive attention.
15. Serving as the single point of contact for project changes, communication of major developments, and issues that may impact the progress of the project.
16. Scheduling and conducting formal reviews with the client during project execution and at completion of the project to confirm acceptance of project deliverables and to obtain feedback about overall team and project performance.
17. Scheduling and conducting formal project reviews in each phase of the project to identify strengths and weaknesses, analyze team performance, and verify the use of standard project management processes.
18. Assuring compliance with the approved organization project management processes.
19. Assuring completion of acceptable project deliverables.
20. Arranging for and assuring an organized close-out process at project completion and managing the hand-off of the project deliverables to the client or receiving organization.

This listing of roles describes what project managers are typically expected to do during their assignment. The actual roles of the project manager will vary by organization, depending on the size of the organization, the types of projects, the organization's enterprise environmental factors and approved processes, and the project management maturity of the organization.

The project team roles are generally assigned through the use of a responsibility assignment matrix. The RAM aligns major project components in the work breakdown structure (WBS) with the project core team members. This defines the responsibilities in connection with the technical aspects of the project.

There are additional roles that the project manager should be aware of, as well. In Dr. Harold Kerzner's book *Project Management: A Systems Approach to Planning, Scheduling, and Controlling, 10th Edition*, Kerzner identifies several team member roles that a project manager may encounter. Team members may demonstrate characteristics defined as *destructive roles* or *supportive roles,* and every project manager should be aware of their existence and potential impact on the behaviors of other team members, the morale of the team, and how these roles affect the project. The project manager's goal is to emphasize the need for, and encourage team members to fulfill, the supportive roles, while discouraging the behaviors associated with destructive roles. An awareness of these roles is important, and although these destructive roles may be undesirable, they are sometimes necessary. These roles may result in some conflict among team members, and possibly between a team member and the project manager. Conflict, when managed effectively, can lead to breakthroughs, new ideas, and greater respect and trust within the project team. Project managers should develop techniques to manage conflict effectively and should understand the benefits or positive side of conflict.

Table 9.1 describes the team member roles that a project manager may encounter as team members are acquired and during the execution of the project activities.

It is clear by the descriptions of these destructive roles that they would be undesirable on any project team. By taking the

Table 9.1 Destructive Team Members

The Aggressor	Openly and continuously criticizes team members, challenges ideas, deflates egos, and tries to eliminate innovation
The Dominator	Manipulates the team, seeks out weaknesses, and tries to take over
The Devil's Advocate	Finds fault in everything and challenges any idea
The Topic Jumper	Switches from one idea to another, creates imbalance and inability to focus
The Recognition Seeker	Always argues for his or her position, attempts to take credit for successes, thinks he or she has the best ideas
The Withdrawer	Does not participate, withholds information, and does not get involved in team discussions and activities
The Blocker	Provides multiple reasons why ideas won't work

Source: Harold Kerzner, *Project Management: A Systems Approach to Planning, Scheduling, and Controlling*, 9th ed. (Hoboken, NJ: Wiley, 2006), pp. 180–181.

time to introduce and explain these roles during the project kickoff meeting and creating an awareness about their negative impact on the project, the project manager can set expectations about what behaviors should not be displayed during the project life cycle. It is important to note that although the destructive roles may be undesirable in most situations, there is actually some benefit to be found when these roles are being engaged. Some conflict may be helpful in the project environment and could uncover new and better ideas for solving problems or increasing the probability for success. Don't use an iron fist approach and set rules and expectations that are totally inflexible. Allow some open discussion and push back. Explaining

Table 9.2 Supportive Project Team Member Roles

The Initiator	Looks for new ideas, uses phrases like, "Let's try this!" or "I'm sure we can come up with a solution if we work together"
The Information Seeker	Tries to become more informed, looks for resources and supportive data. Offers to research for the benefit of the team
Information Givers	Shares what they know, increases the knowledge of the team
The Encouragers	Shows visible support for other people's ideas. Uses phrases like, "That's a great idea," or, "I can support what you are suggesting"
The Clarifier	Helps make sure that everyone understands and issue or a decision
The Harmonizer	Creates a unified feeling among the team
The Gate Keeper	Ensures that all information is relevant and the team stays focused on the issue at hand

Source: Harold Kerzner, *Project Management: A Systems Approach to Planning, Scheduling, and Controlling*, 9th ed. (Hoboken, NJ: Wiley, 2006), pp. 180–181.

what is not desired is only a part of the expectation setting process. The project manager should, upon reviewing the undesired destructive roles, immediately focus on the supportive roles, as highlighted in Table 9.2.

Working with a team that is actively displaying and practicing the supportive roles described in Table 9.2 will significantly increase the chances for project success. By explaining these roles at project start-up, the project manager effectively sets expectations for overall performance and proactively encourages the team to establish an environment that will make the work of the project more enjoyable and possibly fun.

The key here is for the project manager, as the leader, to prevent the destructive roles from developing by actively displaying the supportive roles and acknowledging team members when they display the desired characteristics of the supportive roles. Knowledge of conflict management techniques will also create opportunities to use the behaviors associated with destructive roles to develop better, more effective methods for managing the project and the team.

PROJECT PLAN ACCELERATOR

Using information from this chapter and the *PMBOK® Guide* for additional reference, make a note of any specific planning items that should become components of your project plan. Describe the component and develop questions around the component that will enable you to effectively use the component in the plan.

Establishing Roles and Responsibilities

The project kick-off meeting provides an excellent opportunity for discussing and agreeing upon project team member roles and responsibilities. This important part of the project planning process should not be overlooked or treated lightly. To ensure that your projects needs have been addressed effectively and the right people are assigned to the right activities:

- Review the roles and responsibilities of the project manager that have been established in your organization. Determine if there are any gaps or ambiguous descriptions. Take the necessary steps to adjust these roles. Obtain sponsor support as needed.
- Assess the project team members and determine their strengths in terms of technical and interpersonal skills. One on one discussions would be a great way to obtain this information.

- Develop a set of expectations to be communicated to the project team. Make sure your expectations are clear and intentionally set. Do not make assumptions about expectations and how they are communicated

Consider how the five process groups and the knowledge areas can be used to define project roles and responsibilities. Refer to the planning processes – Initiating, planning, executing, monitoring and controlling to determine who would be involved in an activity and who would be responsible for its completion. Refer to the inputs, tools and techniques, and outputs provided in each chapter of the *PMBOK® Guide 3rd or 4th Edition.* The 4th Edition provides flow charts that effectively illustrate how these process interactions work. It is also important for each team member to be aware of who their specific internal project customers are. Refer to the Project Customer supplier model and emphasize the importance of knowing who will receive each internal project deliverable.

Notes:____

____

____

____

____

____

____

____

____

____

____

____

Project Plan Accelerator

PMBOK® Guide Component / Tool / Technique / Process	*Action Plan for Implementation*	*Person Responsible*	*Date Required*
Human resource planning	Determine the necessary resources for the project.		
Acquiring project team	Identify organizational process assets that are associated with acquiring the project team.		
Developing the project team	Determine the techniques that can be used to create a high-performing team. Define team and individual expectations. Schedule team building activities. Plan recognition events and schedule them throughout the project life cycle.		
Managing the project team	Communicate performance requirements. Determine how team performance will be managed. Develop a process for communicating project performance. Develop skills to manage conflicts. Identify typical conflicts that may occur. Develop response plans for these conflicts.		
Leadership Development	Determine effective methods for motivating the project team. Assess leadership skills and develop a plan for personal improvement.		

Chapter Ten

Risk Management—A Project Imperative

Risk management is one of the most important project planning factors. It is often talked about but is not always included in the actual project management plan. Most project managers would agree that potential risk situations should be identified and assessed at the start of the project planning process, and also throughout the project life cycle. Many organizations see value in identifying and managing risk but fall short of developing risk planning processes because of the perceived lengthy time required to develop a plan and then the perceived effort associated with implementing the plan. There is an underlying notion that risk management costs more to implement than the actual savings that may be realized. In many organizations, risk management is given cursory attention and then basically is handled in a "let's wait and see what happens" type of approach. This often results in disaster and a significant amount of additional work to recover from the risk event.

Risk management should be scaled to meet the needs of the project and to address the specific risk tolerances of the stakeholders and especially the organization sponsoring the project. The core planning process group in the *PMBOK® Guide 3rd and 4th Edition* includes risk management planning. It is an important factor during planning because it has a direct connection to activity duration estimating, cost estimating, procurement management, contract development and scope management, In fact, risk is something that can be easily associated with each project management knowledge area. The entire risk management process includes planning for risk, risk identification, qualitative risk analysis, quantitative risk analysis, risk response planning, and risk monitoring and control. These six processes basically describe the outline of a risk management plan and are arranged in a logical sequence that can easily be explained to project team members and stakeholders. It is important to note that the *PMBOK® Guide* emphasizes the linkages between the five process groups and the nine knowledge areas. The point is to clearly show that the entire project planning process is a combination of activities and processes from all knowledge areas. As an example, during the scheduling process the project team members consider factors that may impact the completion of an activity and add an appropriate contingency reserve to provide a buffer intended to absorb potential risk events. Most of the risk management processes are associated with the *planning process group,* and risk monitoring and control is found as a subprocess under the *monitoring and controlling process group*.

Risk management certainly is, at least it should be, a project planning imperative, but it can become difficult to communicate the value of risk management to key stakeholders and even more difficult to actually implement. During my research about risk management I read several articles that link risk management to disaster recovery planning. A consistent message within each of these articles is this simple statement "If you don't practice the plan, you don't have a plan." That is very true about risk management. The team should understand why the risk management plan is necessary and also understand how to implement the plan. I believe that any project where human life may be at risk must have a risk management plan—not just safety plans and procedures, but actual detailed plans that will minimize the probability of unfortunate events. These projects include constructing large buildings, space shuttle launches, offshore oil rig construction, pyrotechnic projects, and tunnel construction, to name a few. These types of projects utilize highly skilled experts with sophisticated tools and equipment, and there is no doubt that they take risk very seriously.

There are many projects that are not as complex as those just mentioned, but the need for a risk management plan is very much justified. People generally associate risk with negative events and undesirable occurrences and avoid discussing these situations for fear of being perceived as pessimists or as overly cautious resistors to progress. The idea is to look at risk with a positive mindset. Thinking about risk is actually good for the project and for the organization and can lead to great ideas and innovation about preventing negative occurrences while promoting and enhancing positive situations.

I have been to many seminars and discussions about risk management, and I believe the best strategy is to engage the project team in risk management planning at the early stage of the project—preferably during the project kickoff meeting. (You should have noticed by now that the kick-off meeting is a key element of effective planning). I will focus on the qualitative risk analysis process to remain within the scope of this book. The topic of risk management can certainly fill volumes, and the tools and techniques can be very sophisticated and complex. They require a significant amount of information and explanation. The assumption here is that any project requires some type of risk management process, and therefore a simplified approach is provided in this chapter.

In the February 1998 issue of *PM Network* magazine distributed by PMI, an article written by Paula Martin and Karen Tate, PMP, described what I believe to be a straightforward and easily communicated process for managing risk using a qualitative analysis approach. The article, entitled "Team Based Risk Assessment—Turning Naysayers and Saboteurs into Supporters," describes a process that most project teams, especially those working on small to midsize projects with a schedule duration of under a year will find easy and very effective. The process minimizes the administrative burden associated with detailed risk planning documents and processes. Team-based risk assessment is a classic example of how to bring the *PMBOK® Guide* to life by providing a process that takes the key elements described in the *PMBOK® Guide* and applying them to an actual project environment.

The basic process is described as follows:

1. *Make sure the team has been provided with a detailed scope statement and each member fully understands the complexity of the project.* The scope statement and a well-developed work breakdown structure will facilitate the risk management process.
2. *The team identifies the major deliverables associated with the project.* This step provides a description of what specifically will be provided in terms of tangible items to meet the client's needs to achieve the project objectives.
3. *The team uses typical identification processes to list the potential risks associated with each deliverable.* These risks are documented on a chart associated with each deliverable. Project phases can also be used to identify

risks by identifying the deliverables associated with each project phase. Risk identification includes:

- **Brainstorming**—a group discussion in which a multitude of ideas are offered regardless of whether they fit a preconceived image, in order to generate creative solutions.
- **Nominal group technique**—a face-to-face group judgment technique in which participants are provided with a problem or a situation that needs attention. The team generates ideas silently, in writing and anonymously. Responses are collected by a facilitor and posted for review.
- **Delphi technique**—a decision-making process that uses the opinions of experts, generally anonymous, gathered on a dispersed or face-to-face basis, and the guidance and direction of a facilitator to reach either group consensus or a clear definition of alternatives. This technique is used, in part, to minimize bias or influence between subject matter experts.
- **SWOT analysis**—a review of the strengths, weaknesses, opportunities, and threats associated with a project or an organization's goals (see Chapter Three).

4. *The risk events identified by the team are assessed for impact.* This process utilizes the technique known as the probability/impact matrix. The team develops a risk rating by using expert judgment to estimate the probability that the risk event will occur and the impact if it does occur. Risk probability and impact can be determined using a high, medium, low scale (ordinal scale) or a numerical scale (cardinal scale), where high risk may be translated into scores from 8 to 10, medium risks are translated as having a value of 4 to 7, and low risks are rated as 1 to 3.

The probability and impact are multiplied together to produce a risk rating. The higher the risk rating, the greater the need for attention. This process helps the team to prioritize each risk and determine the level of effort and the resources that will be needed to address and resolve the risk situation. Urgency is another factor that should be considered. Some risks that are lower in risk rating value may be more urgent in terms of when they may occur, and in some cases, should be addressed before higher rated risks.

5. *The team assesses the risks.* Team members place the risk events in priority order based on rating and urgency, discuss the reasons for each risk, and begins to plan responses.
6. *The team determines the possible responses or countermeasures for each risk event.* During this process the team discusses possible solutions and or preventive measures. The goal is to generate options for consideration. More effective decisions can be made when more information and alternatives are available.

This process is very easy to use and can greatly improve the overall project management plan. Easy does not mean ineffective, This is a very useful risk management process and can be used in any project phase or at any level of the work breakdown structure. The process also encourages team interaction and helps the team to gain a greater understanding of the project scope and a much better understanding of the value of a risk management process. A by-product of this process is the development of a list of documented risk events that can be added to the organization's lessons-learned files. Continued additions to this list will improve the overall organizational project management process, minimize the occurrence of several potential risk events, and effectively communicate the importance of a risk management plan to all project stakeholders.

Risk Management Lessons Learned

Professional project managers are aware of the value of documenting and sharing lessons learned. Most of the process outputs in the *PMBOK® Guide* include "lessons learned documentation" and experienced and effective leaders continuously add to the ever expanding list of project lessons learned. The following is just a small sample of the knowledge shared by conscientious project professionals:

- Set priorities and revisit them regularly. Things change. What is true today may not have any relevance in the not to distant future.
- Continue to learn. Education never stops.
- Surround yourself with great people. You personally will not have all of the answers. Create teams of people who will seek out the answers and solutions.
- Whenever possible, listen first, decide later. Obtain as much intelligence as possible. Consider other viewpoints.
- Keep your commitments.
- Establish a clear vision and communicate goals and objectives.
- Remain consistent. Don't change your values and your message based on your audience.
- Demonstrate loyalty to your team members and employees.
- Retain your agility and adaptability. Learn new skills and technologies.
- Coach, don't control, your team.
- Create an environment where people trust their leaders.
- Narrow the knowledge gap between worker and management. Encourage workers to assume greater responsibility by empowering them.
- Assess the skills of the people on your team. Know each team member's capabilities.

- Sometimes the best way to lead is to get up and do it.
- Sometimes you have to change the rules.
- First impressions can be deceiving.
- Hope is not a plan.
- Leaders are the ones in the "loop" (stay informed).
- There is no such thing as too much information. Err on the side of giving too much information (I know of several people who would disagree with this one!).
- Remain visible to your organization's people. Talk to them. People appreciate visits from their leaders. It establishes a "connective" environment.

IDENTIFYING PROJECT RISKS—RISK CATEGORIES

The *PMBOK® Guide* refers to the use of a Risk Breakdown Structure (RBS) to assist in the identification of project risk situations. Breaking a project into categories such as technical, external, organizational and project management can assist in further defining project risk events. The knowledge areas of the *PMBOK® Guide* also provide categories that may be used to identify risk situations. Here are some examples:

- Scope management—Risks may include scope creep or failure to define the scope in sufficient detail.
- Time management—Risks may include schedule slippage, lack of resources, and optimistic estimates.
- Human resource management—Risks may include loss of a key team member, conflict between functional groups, and incapable leadership.

Table 10.1 indicates how risk categories can be defined and then further elaborated. Templates such as this are easy to create and provide a framework for project teams to build on.

Table 10.1 Template for Defining Risk Categories

Risk Category	*Sources of Risk*
Technical—evolving design, reliability, operability, maintainability	Examples of sources of risk: Physical properties of the deliverable—size shape, weight etc. Changing Requirements—approved requirements are changes frequently and the changes may cause additional problems Material Properties—density, strength Unstable Technology—unpredictable outcomes Testing Untested Technology—the actual testing process could be a risk Modeling System Complexity—potential for missing key steps or components Integration—failure to ensure design compatibility for multiple components Design—unproven, faulty, no design review Safety—potential for injury during technical testing
Program / Organizational—processes for obtaining resources, enterprise environmental factors, organizational process assets	Material Availability—material may not be available when needed Contractor Stability—potential for a contractor to fail due to financial issues Personnel Availability—critical resources are not available Regulatory changes—government changes regulations Personal Skills—lack of qualified people Organizational Process—faulty or incomplete processes Security processes—missing or incomplete processes Communications processes—inefficient communications techniques, faulty equipment or infrastructure

(continued)

Table 10.1 *(Continued)*

Risk Category	*Sources of Risk*
Supportability—maintaining systems, operating procedures	Product reliability—products fails to meet specifications Training—training is insufficient System safety—safety procedures are not in place Documentation—documentation is missing, incomplete, inaccurate Technical data—data is incorrect, corrupted, lost Interoperability—components do not function with other systems Transportability—inability to move the technology from one system to another
Cost—limited budgets, estimating processes, constraints	Administrative rates—excessive cost for administrative support Overhead costs—additional costs for services or indirect costs Estimating errors—fault data, padding, inaccurate records Cost of quality—repairs, rework, scrap Reliability of estimating resources—techniques used to estimate resources are unreliable
Schedule—estimating processes, reliability of planning processes, methods and procedures	Estimating errors—functional managers provide overly optimistic estimates Number of critical path items—multiple critical paths, large number of critical path items and lack of skilled resources Degree of concurrency—number of items in progress concurrently Unrealistic schedule baseline—schedule based on unreliable estimates, forced end date, contractual obligations

Risk management should be included in any project, regardless of a size. More complicated projects will require more detailed analysis using quantitative techniques such as sensitivity analysis, Monte Carlo simulation, and decision tree analysis: The following is a brief description of each of these techniques

Sensitivity analysis. This compares the relative importance of variables on project outcomes. This type of analysis examines the extent to which the uncertainty of each project element affects the objective being measured when all other uncertain elements are held to their baseline values. The use of a tornado diagram is common in sensitivity analysis. The technique arranges in priority order those items that have the greatest range of impact either positive or negative when compared to the baseline. The image produced resembles the shape of a tornado.

Monte Carlo simulation. This method estimates possible outcomes using a set of random variables and running simulations of the process several times and then observing the outcomes.

Decision tree analysis. A decision tree is a graph of decisions and their possible consequences, (including resource costs and risks) used to create a plan to reach a goal. The decision tree diagram includes the features described in Table 10.2.

These techniques require additional information, data, and specialized tools to produce the desired results. The project team and project manager must assess the appropriate tools to meet the needs of the project and then determine how to implement and manage the tools and the process selected.

One more interesting technique, which is basically a qualitative approach, is a very creative method. I am not sure where

Table 10.2 Features of a Decision Tree

Decision Definition	*Decision Node*	*Chance Node*	*Net Path Value*
What is the decision to be made? Build or upgrade? Purchase or fabricate? Test or not test?	What are the possible choices and the costs of the choice?	What are the possible outcomes of the decision, generally in terms of a positive side and a negative side? The probability of each side is determined, including the reward if the outcome actually occurs. The probability of each branch must equal 100 percent when added together.	What is the payoff minus the costs? This information is then used to calculate the expected monetary value (EMV) of the decision.

this idea was generated, but it makes sense for use on many small to midsize projects:

Step 1. Provide the team with a detailed scope statement. Make sure the team understands the objectives of the project. Everyone on the team should fully understand the purpose of the project and the deliverables to be produced.

Step 2. Ask the team to think into the future and to imagine the project has failed.

Step 3. Then ask the team to list all of the reasons why the project failed.

Step 4. When the team has exhausted all potential risk reasons, ask the team to return their thinking and focus to the present.

Step 5. Finally, ask the team to develop plans to address the listed risks and make sure that everything on the list they created will not happen.

This approach is another way to gain team buy-in and can be very useful in developing interest in risk management, while increasing the team's knowledge about the project.

RISK MANAGEMENT—CYA STRATEGY

There is no shortage of acronyms in the business environment and especially in the project management world. Some acronyms are common across all industries and require no additional explanation. I will make an assumption here that the acronym mentioned here is one that can stand alone.

There will be times when a project manager or team faces a situation in which, despite overwhelming evidence and supporting detail to avoid a high risk project or a bad decision, a directive will be given to plan and execute the project anyway. In these cases, a strategy for action will be required. The following information may be useful if you are faced an assignment that could be described as a "set up for failure". (This would never happen! Or would it?) In these situations, the project manager may not have a choice and must move forward, even though the chances for success seem nearly impossible. This strategy may seem a bit unusual, but I am sure many project managers have been placed in this uncomfortable predicament.

Imagine this situation: You are assigned to manage a project that has been determined by the decision makers of your organization (sponsors or management) to be necessary and or beneficial to your organization. One or more of the following may apply to this project:

- The project has significant risks and an unrealistic imposed completion date.
- The project includes new technology that has not been previously implemented and a contractual completion date has been determined, with severe penalties for missed milestones. The contract was negotiated without the input of a project manager or functional managers. All decisions were made at the executive level with no field input.
- The sponsor is aware of the risks, has seen a detailed risk analysis, and the probability of success is low, but has decided to move forward due to what are perceived to be potential benefits (whether real or imagined). The sponsor is a real risk taker and is also very well connected politically. It may be a *pet* project (a project that actually has no direct connection to the organization's strategic direction).
- The project has been turned down or in some way avoided by all of the "seasoned" and experienced project managers in your organization, who conveniently are far too busy to accept it (or they have political contacts that they are using for a shield).
- All of the above.

The following is a list of project manager activities that could or should be considered in this type of situation. There may be many more actions to add to this list. Use your imagination.

It is absolutely critical for the project manager to define the expectations of the sponsor or executive management. Make sure expectations are clearly stated. Use diplomacy, but obtain and document these expectations. Be persistent. Define and obtain agreement about your personal expectations of the sponsor or manager.

Project Actions/Activities:

1. *Meet with project team.* Explain the project in detail (use a well written scope statement, if possible). Do not complain to the team. Stay focused on facts and the key issues. Avoid getting emotional.
2. *Document feedback from the team.* Do not promote complaints. Work with the team to use their expertise and knowledge when discussing the project. Share this input with your sponsor or manager.
3. *Discuss the project with peers.* Obtain lessons learned, use their insight and expertise. Again, don't complain, just obtain data and support.
4. *Develop what-if scenarios.* What are the consequences of not accomplishing the objectives? (best case/worst case) Discuss and identify other potential risks. Document the results. Conduct what-if scenario sessions frequently.
5. *Prepare status reports.* Depending on the situation, provide status updates weekly, daily, or hourly. This depends on the project and on the expectations of management.
6. *Develop a risk reduction process.* It should include:
   Preventive measures—minimizing risk occurrences
   Mitigation plans—actions that will reduce risk probability and impact
   Work-arounds—temporary responses to risk situations
7. *Keep a daily project log.* Include details such as weather conditions, daily emergencies, extra hours worked, environment issues, out of the ordinary occurrences.
8. *Prepare management summaries.* Provide them daily, weekly or monthly, depending on the situation. Good judgment is a key factor here.

9. *File all documentation, including e-mail.* Keep it easily accessible. You may need this so keep your filing system organized.
10. *Establish a vendor management process to ensure commitments are met.* Set expectations with all vendors and suppliers.
11. *Establish a clearly defined communications process for all stakeholders.* Minimize the possibility of miscommunicating information.
12. *Record all project work time.* Include nonpaid overtime and weekends. Document all of your project work time.
13. *Prepare a detailed project review at the end of each phase and at project completion.* Include an executive summary.
14. *Obtain benchmarking information from other industry experts and consultants.* Having supporting data from reliable sources and unbiased sources will be helpful during reviews and audits and especially during performance reviews.
15. *Do not use warning phrases.* Comments like, "I told you so," or, "you were warned," can really become career damagers.

Regardless of technique used, a risk management process is essential for project success. The team should review techniques that have been used within the organization to determine the best, most effective, and most efficient process for the type of project or situation. Engage project team members in risk management at the start of the project, and make it a point to emphasize its importance throughout the project life cycle.

PROJECT PLAN ACCELERATOR

Using information from this chapter and the *PMBOK® Guide* for additional reference, make a note of any specific planning items that should become components of your project plan. Describe the component and develop questions around the component that will enable you to effectively use the component in the plan. The *PMBOK® Guide* is a resource and can be a great memory jogger during the planning process. Keep it handy for reference.

Risk Management

A risk management process is essential for project success and should be included in any project plan regardless of size or complexity. Take the time to "sell" risk management to your team and project stakeholders. A positive, upfront approach to project risk will reap huge benefits later. The following actions will assist in the development of a risk strategy and process.

- Analyze the culture of the organization as it relates to risk. What is the organization's tolerance for risk?
- Obtain information from other projects about how risks have been managed or check existing organizational processes for information that may facilitate the planning process. Are there any standards in place? Check with other project managers. There may be techniques being used that have a direct connection to risk management.
- What type of analysis is most appropriate? Qualitative or quantitative? This depends on the type of project.
- Communicate the importance of risk management to the project team. Provide examples of projects that became troubled due to a lack of attention to risk.
- Emphasize the importance of discussing possible risks on a regular basis during project status meetings. Discuss potential "show stoppers." Create a "contingency prepped" team.

Consider how the five process groups and the knowledge areas can be used during the development and implementation of the risk management plan. Consider the methods that may be used to identify risks, rank the risks, determine responses to the risks, and monitor and control new risks. Each knowledge area of the *PMBOK® Guide* can be considered a category for risk. Use these knowledge areas to generate discussions about risk.

Notes:____

____

____

____

____

____

____

____

____

____

____

____

Project Plan Accelerator

PMBOK® Guide Component / Tool / Technique / Process	*Action Plan for Implementation*	*Person Responsible*	*Date Required*
Risk management plan	Determine who will be responsible for developing the risk management plan or contributing to the plan. Determine the risk tolerance of your organization (Risk Averse, Risk Neutral, Risk Seeker). Review the enterprise environmental factors associated with your organization.		
Risk identification	Determine the most appropriate techniques for risk identification to be used on the project: Brainstorming Nominal group technique Delphi technique SWOT analysis Identify, rate, and rank the project risks. Use a probability/ impact matrix and team based risk assessment. Create a risk register. Educate and train the project team about these techniques.		

(*continued*)

Project Plan Accelerator (*Continued*)

PMBOK® Guide Component / Tool / Technique / Process	*Action Plan for Implementation*	*Person Responsible*	*Date Required*
Qualitative analysis Quantitative analysis	Create a risk breakdown structure. Initiate a team based risk assessment for the project. Determine probability and impact of each risk identified. Develop risk responses. Quantitative Analysis techniques include: Monte Carlo Simulation Decision Tree Analysis Sensitivity Analysis Determine which approach qualitative or quantitative or if both should be used to analyze and prioritize risks. Review the tools and techniques used for these processes. Conduct training sessions for the team or obtain experience individuals.		
Risk response planning	Determine the most appropriate responses to each risk identified: Avoid—find an alternative. Transfer—hand off to a qualified expert. Mitigate—reduce probability and impact. Exploit—create an environment where a positive risk event can likely occur.		

	Enhance—obtain the greatest benefit from the positive event. Share—include other organizational entities in positive risk events. Accept—accept the risk event and prepare an appropriate back up plan or contingency in case the risk event occurs. It is important to ensure that detailed explanations for each response to an identified risk are documented by the individuals involved. This information will become useful lessons learned for future projects.		
Risk monitoring and control	Assess responses to the identified risks. Determine if the response is appropriate and provides the most effective solution. Review primary risks—obvious risks that may have a serious negative impact on the project. Identify secondary risks—risks that may occur as a result of the decisions made to address primary risks. Identify new risks—risks that are not identified during the planning process. Monitor residual risks—risks that have been reduced but have not been eliminated. Manage the cumulative effect of risks (the effect experienced when many risks appear to be impacting a project at the same time, creating the appearance of a great or very severe threat).		

Chapter Eleven

A *PMBOK®* *Guide* Strategy for Success

Understanding the project environment is a key factor for success. The *PMBOK® Guide* identifies and emphasizes the need to become aware of the cultural and social environment of the business and the project, the international and political environment, and the physical environment that may affect the project outcome. Significant emphasis on general management skills such as planning, organizing, and staffing, along with a need to become familiar with the operating environment, is also critical. An understanding of financial management and accounting, purchasing and procurement, and the legal issues associated with managing an organization are also a major part of the project environment.

A review of Chapter 4 of the *PMBOK® Guide, 3rd Edition and 4th Edition*, "Integration Management," which was added to the *PMBOK®* in 1996, will indicate that integration is the key to a successful project and that the project manager must combine and coordinate all of the elements of the other eight knowledge areas into a high-level view of the entire project process. It is not an all-inclusive view, (many of the details associated with integration are discussed in other chapters) but it is intended to demonstrate the **systems approach** to managing a project.

Each knowledge area is part of a system. There are no stand-alone components. Every knowledge area is equally important. One missing piece could easily send the project into chaos. In this author's opinion, the "Integration Management" chapter should have been placed as the final chapter of the *PMBOK® Guide*. The rationale behind this opinion is that each of the knowledge areas is explained in detail *after* the "Integration Management" chapter. Key terms used to explain each knowledge area are used in the "Integration Management" chapter, which may require the reader to read other chapters and then come back to the "Integration Management" chapter. (I believe there is an assumption that a reader of the *PMBOK® Guide* already has an in-depth knowledge of project management but discussions with hundreds of project managers indicates this is not the case.) A brief introduction about the relationships of all knowledge areas and an explanation about how all processes are integrated should be inserted where Chapter 4 currently exists. The project context and the major process groups are explained in Chapters 1 to 3, but some type of "systems" view would be helpful to explain how the knowledge areas are integrated and support each other through the process groups. The reader can then review each chapter, understand the key elements of each knowledge area, and then conclude the review and or study of the *PMBOK® Guide* by reading the final chapter about integration which would tie all knowledge areas together as a system.

A suggested strategy for reading and using the *PMBOK® Guide 3rd and 4th Edition* is as follows:

1. Read Chapters 1 through 3 in one sitting. These three chapters are linked and provide an explanation of key

terms, as well as an introduction to the process groups found in each knowledge area. You need this information to fully understand how the *PMBOK® Guide* is written and the logic that is used. Remember, it is not *the* body of knowledge of project management, but it is an excellent foundation for understanding, developing, and utilizing formal project management processes and methodologies.

2. Read the *PMBOK® Guide* glossary. This may seem an unusual suggestion, but the glossary does provide an enormous amount of useful information about project management and establishes a common language that most project managers should be familiar with. The terms in the glossary appear frequently throughout the document. This is also a good idea for project managers preparing for the PMP exam. Many of the terms are more clearly defined in the glossary than in the actual chapters.
3. Next, briefly review Chapter 4. Examine the overview chart for integration. There are a few differences between the 3rd and 4th Editions of the *PMBOK® Guide*. The 4th Edition places more emphasis on the actual integration of the knowledge areas and the coordination steps that are used to manage a project and not as much on the details of each of the knowledge areas. Focus on the project charter and review the list of enterprise environmental factors and organizational process assets. These items can have a profound impact on the project planning process. Enterprise environmental factors include such items as the organization's culture, the infrastructure, and the capabilities of the organization's human resources. Organization process assets are associated with internal procedures and processes that have been established such as purchasing, obtaining contractors, and managing

change. The integrated change control process explained in Chapter 4 of the *PMBOK® Guide* is basically a summary of the change control processes found in each of the other chapters. Change control is a critical factor in project planning and execution and gaining an appreciation of how change in one area can have a significant impact in many other areas of the project planning process is extremely important. *The Integration Management* chapter provides the "big picture" view of the project management process. Refer to the integration overview chart frequently as you read about and develop an understanding of each knowledge area. The objective here is to ensure the connection and interdependencies of each knowledge area.

4. Read each succeeding chapter from Chapter 5 through Chapter 12. Read each chapter thoroughly, and make notes or highlight key information and terms. It's best to schedule some buffer time between each chapter. This allows you to reflect on the information presented and prevents the feeling of becoming overwhelmed by the enormous amount of data. There is a very large amount of information provided, and it is important to make connections between the chapter you are currently reading and the previous chapters. Each chapter builds on information from the previous chapters. Once you have completed reading and reviewing Chapter 6, you will notice a significant repetition of terms. This repetition actually provides a means to demonstrate how the knowledge areas and processes are integrated.
5. Upon completion of a chapter, review the overview chart provided within the first three pages of each chapter. These overview charts describe the generally accepted inputs, tools and techniques, and outputs associated with the processes for each knowledge area. Use the

overview charts to test your knowledge. You should be able to explain each input, tool or technique, and output. You should know why each input is important, and its source of origin. You should be able to explain the various tools and techniques and why each one could be used to produce the desired results and how outputs become inputs to other processes. (There is no need to memorize the detailed information in the *PMBOK® Guide* but you should be able to explain reasonably well what each item on the chart is and why it is important.) Remember the *PMBOK® Guide* is a reference document and should be referred to frequently during the project life cycle.

As an example, if you review the inputs to 5.1 of "Project Scope Management—Scope Planning in the 3rd Edition or 5.2 Define Scope in the 4th Edition," you will see the following inputs: enterprise environmental factors, organizational process assets, and project charter, or tools and techniques: expert judgment and product analysis. You should be able to explain each of these inputs or tools and techniques and why they are needed (or not needed) in the overall scope planning process.

You will notice that the input *enterprise environmental factors* is included in many of the processes explained in the *PMBOK® Guide*. These factors are associated with organizational culture, industry standards that may affect your organization or the project you are assigned to, organizational infrastructure and support systems, and stakeholder risk tolerances. These environmental factors are key items to consider during the project planning process. They will impact how the project manager, team, and project sponsor make decisions and determine what tools or techniques may be selected to achieve the desired project results. *Organizational process assets* is also found as an input in many process groups. These process assets are the formal and informal policies, processes, and standards used by an organization for general operation. Project managers and teams should become aware of these process assets and how they may impact project planning and decisions during implementation. Process assets include safety and health policies, templates for planning, change control procedures, and financial controls.

6. Continue to review each succeeding chapter and then make a connection with each previous chapter by reviewing and comparing information. Look for new terms, new tools and techniques, and other processes that have been introduced. Make sure you are aware of the purpose of these items and how they relate to your projects. Refer to the glossary for assistance and also use other project management resources for further, more detailed explanations. Look for items that have been discussed in previous chapters. Continually ask yourself, "How can I apply this process to my current project?" Why isn't this process applicable to my projects? How can I adapt this process to work within my project? This continuous review will drive you toward a deeper understanding of the *PMBOK® Guide and the larger project management body of knowledge.* It will enable you to manage your team more effectively and produce better, more acceptable project plans for your stakeholders.

The *PMBOK® Guide* has probably been analyzed by countless project managers, and I am sure there is a huge range of opinions, from, "I can't use this, it's too inflexible," to, "My organization is fully *PMBOK®* compliant." The key is common sense. After all, it does say *guide*. I think the *PMBOK® Guide* is a great source of information, and it will continue to evolve and improve. I see it as a tool, and like any tool, you have to become familiar with it and how it can be used. Most tools have more than one function and the *PMBOK® Guide* is no different. As you use it, you will find that it can provide you with far more than the basics of project management and can actually sharpen other tools in your project manager's tool box.

PROJECT PLAN ACCELERATOR

Using information from this chapter and the *PMBOK® Guide* for additional reference, make a note of any specific planning items that should become components of your project plan. Describe the component and develop questions around the component that will enable you to effectively use the component in the plan.

The *PMBOK® Guide* as a Planning Tool

The *PMBOK® Guide* is a great tool for planning projects. The 3rd or 4th Edition can be used to greatly improve any project planning process and help to establish a workable project management methodology. As I have stated several times common sense is the best approach to project planning and proper use of the tools will certainly improve the overall quality of any plan or methodology. To use the *PMBOK® Guide* effectively the following steps should be considered:

- Review the five major process groups and how they integrate the entire project management process. Identify the similarities between these processes and the processes in use within your organization. Look for areas where you can improve your process or where process gaps may exist.

- Develop a *PMBOK*® *Guide* learning plan for your organization. How can the information provided in the document improve how your organization manages projects? Consider a gradual approach if your plan is to create a formalized enterprise process.
- Read the document frequently. You will probably find something that you didn't read before. (It's easy to miss something.)
- Review each knowledge area and analyze the inputs, tools and techniques, and outputs. Discuss these items with your team. Determine which of the inputs, tools and techniques, and outputs are most applicable to your projects.

Consider the types of projects you are likely to be assigned to. Use the *PMBOK*® *Guide* as a tool. It's not a one-size-fits-all type of tool. Customize your processes to meet your organizational needs. The *PMBOK*® *Guide* is a collection of generally accepted best practices and will help you to create your own set of best practices.

Notes:____

Porject Plan Accelerator

PMBOK® Guide Component / Tool / Technique / Process	*Action Plan for Implementation*	*Person Responsible*	*Date Required*
Process group interaction	Review how the process groups interact in the project life cycle. Review your organization's project life cycle and compare it with the process interactions. Look for overlapping processes and how knowledge areas are integrated in each phase.		
Initiating	Define the specific initiating processes that are used for your projects, at the project level and in each phase. Refer to the Project Management Process Interactions described in the *PMBOK® Guide* and compare with organizational processes. Identify areas for improvement.		
Planning	Review the planning process group. You will notice that there is significant integration of knowledge area processes within the planning process group. Define the project planning process for your project or for your organization. Review the planning processes described in the *PMBOK® Guide*. Determine which processes are necessary for the assigned project.		

(continued)

Porject Plan Accelerator (*Continued*)

PMBOK® Guide Component / Tool / Technique / Process	*Action Plan for Implementation*	*Person Responsible*	*Date Required*
Executing	Review the executing processes and determine which processes are applicable to your projects. Execution includes performing quality assurance, acquiring the team, developing the team, and requesting seller responses. Identify who is responsible for these activities. Create a project execution strategy.		
Monitoring and controlling	Establish mutually agreed upon monitoring and control procedures. This includes escalation plans, integrated change control, and observing project performance. Determine which subsidiary plans are needed for your project (subsidiary plans are detailed support plans for specific areas). Develop the following plans as needed: Scope control plan Cost control plan Schedule control plan Quality control plan Risk monitoring and control plan Safety plan Cutover plan Disconnect Plan Transition plan		

	Assign responsibility for development of the required subsidiary plans. (Avoid unnecessary administrative work. Focus on what is needed and coach team members to develop effective but simplified plans whenever possible.)		
Closing	Ensure that you have a process for closing the project. Identify the acceptance criteria for all deliverables. Create an organized approach to ensure a complete close-out and acceptance by the customer. Create a close-out work breakdown structure. Develop a customer satisfaction interview process. Schedule a project review and lessons-learned session. Plan for team recognition. Prepare all documents for handoff. Prepare a final executive report. Review and verify completion of all contractual deliverables. Perform a vendor/supplier analysis.		
Communications management	Determine stakeholder information and communications requirements and develop a method for information distribution. Determine what information will be communicated to each stakeholder. Develop a method to ensure that information is received and understood by all stakeholders (feedback loop). Establish a process to record project communications for future reference (back up files for communications). Ensure that targeted stakeholders receive and act upon information in a timely manner (set expectations).		

Chapter Twelve

Rapid Knowledge Development (RKD) For Project Managers: Tips, Tools, and Techniques

The *PMBOK® Guide* provides project managers with an organized view of the higher, broader, and much more complex project management body of knowledge. As previously mentioned, the nine knowledge areas are integrated by using the five major process groups to create a logical arrangement of tasks and activities and to develop the appropriate subsidiary plans that will support the project plan. The knowledge areas provide details about the specific inputs, tools and techniques, and outputs that may be used to progressively elaborate the project plan and to integrate the process groups (initiating, planning, executing, monitoring and controlling, and closing). The process groups also create an awareness among the project manager and team that many project related planning activities are developed concurrently. The integration of the knowledge areas through the major process groups ensures a connection and systems approach to create a plan that is fully defined and articulated to the required level of detail. The defined and continued emphasis on the relationship between the knowledge areas and the five process groups effectively ensures the integration of each project

planning element and supports a systems approach to managing the entire project life cycle. The systems approach will create a project plan and a methodology based on three key factors:
Every element of the project is

- Integrated
- Inter-related
- Inter-dependent

This means that there are no separate, "dangling" elements in the project plan. There is one main set of objectives and all planning components are connected to meet those objectives. Dangling components (components that are not connected to an end task or milestone and appear to be "floating" in the plan) could create risk situations, undesired outcomes, or failure to deliver a key project component. The systems approach will ensure that the entire team fully understands the need to clearly define all project components, and then connect them in a logical and efficient manner to produce a complete, final deliverable. This includes how the team will be organized and the inter-relationships between team members.

An analogy of the systems approach is the familiar tabletop puzzle. A puzzle contains hundreds of pieces (maybe thousands), and each one is equally important. When all of the pieces have been placed in the correct location and all interfaces have been established, the final product (the picture or image) is presented. Just one missing piece or a piece that is not placed correctly will be noticed immediately. This missing or incorrectly placed piece can create a system wide problem, and the deliverable will be noticeably flawed. In most cases, it will not be accepted until the problems are corrected. The familiar *triple constraint* (on time, within budget, and according to specifications) that is generally associated with project plans is a good example of the interrelationships that must be understood and balanced during planning and execution. Each component of the triple constraint is part of the greater system and changes to one element of the triple constraint will have an immediate affect on another element and probably the project as a whole.

The *PMBOK® Guide* is one of the many tools that should be included in that ever-expanding project manager toolbox. The information in the *PMBOK® Guide* is actually a catalog of information for use by project managers and teams to organize the planning process and to determine what tools and techniques should be used to create an appropriate plan for the project. The best practices identified and explained in the *PMBOK® Guide* should be analyzed to determine which practices are needed for the project that is being planned. We know that all projects are unique, and therefore, each plan will be unique. Treating each project plan as unique (even if the project is nearly identical to a previously completed project) helps to minimize complacency and keep the team focused.

A key element not always mentioned in literature about project management is something very simple: common sense. Common-sense project management, explained in simple terms, means this: Use a proven methodology whenever possible, plan the project only to the detail required, minimize administrative tasks, avoid over planning, use the knowledge of the team, verify all assumptions, make a big deal out of freeze dates and the change process, and, as my mentor in project management, Dan Ono, always said, "Go slow now so you can go faster later." Managing a project can be a very challenging assignment and the intent here is not to create a perception that it's a "walk in the park." The emphasis should be on smart, common sense planning and a managerial approach that will motivate, not inhibit the team members and activity performers.

CONNECTING THE *PMBOK® GUIDE* AND *THE* PROJECT MANAGEMENT BODY OF KNOWLEDGE

As discussed in Chapter One of this book, the *PMBOK® Guide* provides the project manager with an organized view of the higher-level, more-comprehensive project management body of knowledge. We know the true body of knowledge of project management is a complex, highly integrated collection of information gathered over hundreds of years. It includes elements of people management, engineering, financial management, time management, leadership, motivation, managing conflicts, diversity, and much more. The actual project management body of knowledge is something that may not ever be truly defined in one document. A good model of what could be considered one example of the project management body of knowledge can be found at the Web site www.npd-solutions.com/bok.html. This Web site is the home of the *New Product Development Body of Knowledge* and is a great example of just how extensive a body of knowledge about a specific discipline can be. This Web site addresses strategic planning, product development, competitive analysis, innovation management, technology deployment, business cases, R&D, managing teams, leadership, and much more. Project managers, especially those who are just beginning to experience the challenges of the profession, will find this Web site extremely useful and a significant source of detailed information. It is just one of the many gateways available to the ever-expanding project management body of knowledge.

THE BODY OF KNOWLEDGE COMPRESSED

In this new century marked by fast-paced and evolving technology, worldwide competition, and an uncertain economic environment, sources of summary information, quick data, and key points about planning and implementation are useful for newly assigned project managers or people who are assigned to manage a project but are primarily involved in operations-type work. Due to the often high-pressure and hectic nature of the project management position, there may be little time available to attend a training program or skills development course about the fundamentals of project management. Time constraints, competitive pressures, the multiproject environment, and impatient sponsors or executives require an accelerated learning curve. There certainly is no substitute for a well-executed project management training curriculum and the experience gained by managing a project, but having a summarized reference of key points, tools, formulas, and processes can be very helpful when managing through the daily challenges experienced from beginning to end of a project cycle.

The following information is a digested view of the key items generally associated with the nine project management knowledge areas and their relationship to the five major process groups. These memory joggers may assist in developing a greater understanding the complexity involved in developing and executing project plans and the importance of using an integrated and coordinated approach. Each knowledge area is summarized and includes a brief description of terms, tools, techniques, and other items that are part of the overall project management planning process. This summary information will also be helpful for project managers who are planning to study for the Project Management Institute PMP® exam.

PROJECT INTEGRATION MANAGEMENT

Integration management is the one area in the *PMBOK® Guide* that I believe needed some extensive additional work in the 3rd Edition to meet its objective of tying all of the knowledge areas and process groups together. I see this knowledge area as the key element that connects all project components and sets a foundation for project success. Integration management is about coordinating, unifying, and managing each major process to successfully complete the project. It's the big picture view, and it requires the skills and knowledge of a project manager who understands the technical elements of project planning as well as the human side. To effectively integrate all of the components of a project plan, the project manager must also be familiar with the skills associated with that of a general manager. The project manager becomes *the* integrator of the major processes, the stakeholders, and all of the subprocesses and activities required to plan, execute, and control the project. In many ways, the project manager position is similar to a CEO. There is a need for leadership, a technical understanding of the major elements of the project, vision, and the ability to bring a work force and management team together to complete the agreed-upon objectives.

The chapter "Integration Management" in the *PMBOK® Guide* 3rd Edition includes the following:

- Develop the project charter
- Develop the Preliminary Project Scope Statement
- Develop project management plan
- Direct and manage project execution
- Monitor and control project work
- Integrate change control
- Close the project

There appeared to be a few gaps in this chapter regarding how all of the parts are actually being integrated and how to view the larger picture. The contributors of the 4th Edition must have noticed the need for some adjustments and have created a more realistic and logical view of Integration management. The 4th Edition includes:

- Develop the project charter
- Develop project management plan
- Direct and manage project execution
- Monitor and control project work
- Perform integrated change control
- Close project or phase

It is important to note that the processes listed in the *PMBOK® Guide* are not necessarily sequential. There will be some degree of overlap between these processes especially the processes of: develop project plan, direct and manage project execution, monitor and control project work, and perform integrated change control.

Integration management should be viewed as the overarching knowledge area, or the umbrella knowledge area under which all of the other knowledge areas exist. For example, in the 3rd Edition the "Integration Management" chapter includes the processes associated with the development of the project charter *(Note: One of the reasons why charters were created was because the project manager was brought on board after the project was approved and was not privileged to the rationale concerning the approval. Therefore, the charter was created to document the assumptions)* and the development of the preliminary scope statement. There is no mention of the finalized scope statement in the process group "Develop Project Management Plan." It seems logical under "Integration Management" to show that scope definition includes the processes for developing the preliminary and the finalized scope statement. These activities will lead to the integration of all elements of the project plan through the five major processes.

I see integration management to include an understanding of the chartering process, the contract terms and conditions when applicable, and the progressively elaborated scope statement.

These items are, in my opinion, the key inputs to the project management plan. There should be no question or debate about the need to understand how enterprise environmental factors and organizational process assets affect decisions about project selection and planning processes, but the integration of each of the processes, subprocesses, and knowledge areas requires a clear scope definition and a clear understanding of how all of the parts come together to achieve project success.

It should be understood that the *PMBOK® Guide* provides an outline of what is commonly included in a project plan and the best practices associated with plan development, but there is a need for the "Integration Management" chapter to provide a summary that is more consistent with the information provided in each of the other eight knowledge areas. Integration management is what project managers do, and this subject should be a reasonably detailed summary of the key elements required to successfully develop a plan, manage a team, control changes, and deliver a completed set of deliverables. The *PMBOK® Guide* 4th Edition address integration management in a sensible manner that actually demonstrates how each knowledge area is connected and provides a more realistic view of how all of the components of the project life cycle work together to achieve the desired results.

Key Terms and Processes Associated with Integration Management

- **Contract.** A legal document that may be remedied in a court of law. It provides the details that constitute an offer, acceptance, and a consideration (something of value) regarding services between two or more entities. Generally, the terms and conditions of a statement of work and a contract will provide the basis for the project charter and scope statement.

- **Historical records.** Archives of information from previous projects. The idea here is to learn from past projects and prevent mistakes and issues from recurring. Use documented lessons learned and the experience of the team to assist in project plan development.
- **Project selection methods.** The project selection process is usually conducted prior to approval of the project charter. These methods include techniques such as net present value, discounted cash flow, payback period, break-even analysis, and other benefit measurement methods. Project sponsors and steering committees should be provided with information about the business value of a project to ensure the most effective use of organizational resources. Mathematical modeling such as decision tree analysis where probability and outcomes are predicted using available financial data and predictions about performance may also be used.
- **Enterprise environmental factors.** These factors will be found as inputs in many processes described in the *PMBOK® Guide*. Examples of these factors are: organizational culture, tolerance for risk, human resources and capability, infrastructure, government regulations, and compliance issues.
- **Organizational process assets.** Consider the policies and procedures defined by your organization and how they can impact the planning of your projects. These might include the hiring process, budget policies, vendor selection process, and so on.
- **Constraints.** The limitations you must operate within. Examples include predetermined project end dates, contractual milestones, funding, and resource availability.
- **Assumptions.** Assumptions are defined as those items that can be considered true, real, or certain for planning

purposes. Consider the assumptions that can be made about the project. Examples of an assumption set: There will be dedicated management support for the project, the resources required for the project will be available, sufficient funding is available, and the appropriate technical knowledge is in place.

- **Project charter.** This document formally authorizes the project to use organizational resources and is used to assign the project manager and describe the project managers level of authority. The project charter documents the business needs, high level risks associated with the project, initial set of stakeholders, and can include a summary schedule and milestones.
- **Stakeholders.** There generally accepted definition of a stakeholder is "anyone directly involved in or in some way impacted either positively or negatively as a result of the project. Consider who the *key* stakeholders are, and also other stakeholders who may view your project as a threat or an obstacle to their project (negative stakeholders). Determine who the negative stakeholders are and what risks they may introduce to the successful completion of your project. Key stakeholders generally include the project manager, sponsor, project team, customer, and end user.
- **Organizational strategies.** Consider the goals and objectives of your organization and how your project impacts or supports them. Make sure you can link your project to the organizational goals.
- **Project management information system.** This is any system or group of systems working together to gather, store, and distribute information about your project.

Examples include time reporting system, accounting system, and project software.

- **Project management plan.** You will be documenting the collection of planning outputs that will guide project execution. The project management plan generally includes the subsidiary plans such as a change control plan, communications plan, scope control plan, and so on. The plan is expected to change as the project is progressively elaborated and as risk situations develop requiring the project team to replan their approach.
- **Project management methodology.** A methodology is defined as a "way of doing something." The organization develops a consistent approach for managing project planning and execution.
- **Corrective and preventive actions.** Corrective actions are those actions reviewed and decided upon by the project team to resolve a problem such as an unacceptable variance or an authorized scope addition. Preventive actions are those actions taken to minimize errors, risks, and other negative factors.
- **Earned value technique.** This is a process that is used to analyze project performance by observing and managing three main factors: Planned Value or BCWS (budgeted cost of work scheduled). It can be translated as the estimated cost of the work that ***should*** be accomplished at the time of measurement, Earned Value or BCWP (budgeted cost of work performed), the estimated cost of the *work* that has been *completed* at the time of measurement, and Actual Cost or ACWP (actual cost of work performed), which is the dollar amount paid for the work that has been completed at the time of measurement. Earned

value analysis is basically an early warning system for your project and will assist in identifying performance trends and managing variances.

- **Work authorization system.** The processes that ensure the right work is done at the right time, in the right sequence, form the work authorization system. This system is designed to prevent work that is not authorized and assures that work is done safely and according to the plan. It may reduce or eliminate many project risk situations.
- **Integrated change control.** The nine knowledge areas are managed in an integrated manner with an understanding that a change in one area can affect any or all of the other knowledge areas. Consider the impact of the change before implementing the change and obtain approval for the change. Establish and refer to project baselines frequently to determine if a change has occurred through comparison of the baseline with actual results. Include in the process a method to determine when to make a change and how to introduce the change to minimize the impact on project performance, ongoing operations or other projects.
- **Change Control Board.** This board is a team or group of designated decision makers (possibly a steering committee designated or empowered to review and determine the value of a change and to approve or deny change requests.
- **Configuration management.** This process will ensure that configuration changes (changes to features, functions, physical characteristics) are managed and approved to prevent or reduce the risks of additional cost, scope changes, or other impacts to the project.
- **Subsidiary plans.** Detailed plans should support the project management plan. They may include the scope

change control plan, schedule change control plan, cost change control plan, and so on.

- **Deliverables.** These are tangible, verifiable work outputs. Projects generally produce several deliverables that will support the final or ultimate objective of the project. Examples include the product of the project, the major components of the product, training manuals, and transfer documents.
- **Administrative closure.** These are the processes associated with closing all open project records, such as financial accounts, work orders, training documents, and change requests.
- **Contract closure.** This includes verification that all conditions of the contract have been fulfilled and each party involved agrees to formal acceptance.
- **Work performance information.** This includes schedule status, cost information, level of quality, percent complete, completed activities, and resource utilization.

A comparison of the 3rd and 4th Editions of the *PMBOK® Guide* will show some changes in the process groups that comprise Integration Management. The 4th Edition provides a more straightforward and more appropriately arranged view of the key elements associated with effective Integration Management.

PROJECT SCOPE MANAGEMENT

The major process groups for scope management in the 3rd Edition of the *PMBOK® Guide* are: scope planning, scope definition, Create WBS, scope verification, and scope control. In the 4th Edition Scope planning has been replaced by "Collect Requirements." Scope Definition is now "Define Scope", Scope

Verification is renamed Verify Scope, and Scope control is now "Control Scope. These are not significant changes but they help improve the overall readability of the document. These major groupings within scope management provide the foundation for planning the project in detail and ensuring that the project team members understand the complexity of the project. It is important to note that, although scope management and other knowledge areas are shown separately, they are actually integrated into the larger project planning process through the five major process groups. Each knowledge area is described separately for learning purposes, and it should not be assumed that the processes described must be followed in the sequences shown.

It is important to review and understand the inputs, tools and techniques, and outputs of each process within each knowledge area. Ask yourself why each input *is* an input and where it originated. Why is it important? Understand what each tool and technique will produce and how each is used. You may not need each tool or technique listed, but it is important to learn about how they are used for possible application in a future project. Tools and techniques produce outputs, and the outputs of each process generally become inputs to other processes. This pattern is consistent throughout the *PMBOK® Guide* and is intended to promote the systems approach to planning, executing, and controlling a project.

Scope Planning

In the 3rd Edition of the *PMBOK® Guide* the key inputs to scope planning are project charter, preliminary project scope statement, and the project management plan. Remember that enterprise environmental factors and organizational process assets are common inputs to many process groups.

Scope planning tools and techniques include expert judgment (generally the functional managers and subject matter experts), templates, and standards. Templates, in many cases, may be obtained by utilizing information from previous projects such as task lists, work breakdown structures, and other project documentation.

The 4th Edition of the *PMBOK® Guide* replaces Scope planning with "Collect Requirements." In this sub process group there is more attention to the requirements process including the use of focus groups, facilitated workshops, and prototypes. This is more closely related to the actual practices in many organizations. A generally accepted process of requirements management includes: Elicitation, Analysis, and Specification. The 4th edition provides a closer alignment to the requirements process and includes a brief description of widely used techniques for gathering requirements. These techniques include:

- Interviewing
- Focus groups
- Group decision making techniques such as Delphi, Nominal Group, and brainstorming

Requirements collection and management also requires a specific process to ensure that the *real* requirements are documented. A typical requirements planning process includes the following steps:

1. Identify the team to be assigned to the requirements process.
2. Document team member roles and responsibilities.
3. Define requirements collecting deliverables and time frames.
4. Define the process and determine which techniques will be used: interviews, meetings, workshops, etc.

5. Outline existing operational needs. Obtain or produce workflow and operations charts and other process artifacts. (Review organizational process assets for requirements management.)
6. Schedule workshops and requirements collection sessions to begin the initial requirements collection process.
7. Schedule requirements review sessions to ensure the collection of the real requirements.
8. Establish a change control process for requirements.
9. Establish a requirements traceability matrix or system for traceability.

According to Ralph R. Young, author of *Effective Requirements Practices,* Addison-Wesley Information Technology Series, attributes of a well written requirement should include the following:

Requirements should be

- Necessary
- Verifiable
- Attainable
- Unambiguous
- Complete
- Consistent
- Traceable
- Concise
- Implementation free
- Include a unique identifier

Scope Definition or Define Scope

The project scope management plan, an output of scope planning in the *PMBOK® Guide* 3rd Edition provides guidance about how the scope is defined, documented, verified, managed, and controlled. Scope definition further elaborates the project and provides more detail for planning purposes. This includes stakeholder analysis, product analysis, and alternatives identification. The 4th edition includes requirements documentation as a major input to defining the scope. The tools and techniques remain basically the same and include the addition of facilitated workshops where stakeholders are brought together to discuss requirements and specific individual needs.

The scope definition process produces the **project scope statement.** The scope statement basically answers the questions: What will be done? Who will do it? When must it be done? Why are we doing it? Where will the work be performed? How much will it cost? A well defined scope statement will create a stronger planning process and facilitate the development of a well organized project plan and committed team.

A note about the scope statement—The scope statement is actually a re-writing of the customer's statement of work in such a language that the project team can understand it and use it to develop the more detailed plans. The PMBOK® Guide does not make this clear, but the project manager prepares the scope statement, not the customer.

Create WBS

The WBS (Work Breakdown Structure) is defined as a deliverable-oriented hierarchical decomposition of the work to be executed by the project team. The WBS (work breakdown structure) is created using the process known as decomposition. The process of decomposition, a *PMBOK® Guide* term, breaks the project down into smaller elements. The term decomposition actually refers to the break down of tissue of a formerly living organism into simpler forms of matter. You can see the analogy.

The *PMBOK® Guide* defines decomposition as the subdivision of project deliverables into smaller, more manageable components. To facilitate the process of creating a WBS Templates may be obtained from previous projects or established standards developed by a PMO (Project Management Office). The WBS breaks down the project into smaller, more manageable components and allows for more effective and accurate time and cost estimating. The WBS is also used to identify project risks and can be a very useful team-building tool.

WBS development is concerned with the grouping of tasks and activities and is not intended to display the sequence of tasks and activities. Avoid attempting to sequence tasks in a WBS and focus on grouping the tasks. The objective is to identify the tasks that must be completed to produce the final project deliverables.

A WBS dictionary is a document that provides the team and other stakeholders with more detail about WBS components and tasks. This can be a separate document included within the project plan. The level of detail and completeness of the WBS dictionary depends on the complexity of the project and the specific needs of the team members assigned to the tasks and activities.

The scope baseline is produced during the process of creating the WBS. The baseline provides the basis for more detailed planning and a more complete view of the project to be implemented.

Scope Verification or Verify Scope

Scope verification is about acceptance of the actual result of the work. Verification is accomplished through inspections, reviews, and audits. It is the actual review of the deliverables produced during the execution of the project plan.

Scope Control or Control Scope

Scope control includes the scope change control system, variance analysis, replanning processes, and configuration management process. An effective scope control plan will help to minimize scope creep (the unauthorized additions to the approved project scope of work). Variance analysis is the main technique used for controlling project scope. The defined and approved scope statement becomes the baseline and the project manager and team review project performance on a regular basis to detect any changes to the baseline.

Configuration management is concerned with managing changes to the features, functions, and physical characteristics of the deliverables of the project. Configuration management is actually a component of the integrated change control process. Additionally, configuration management is associated with more than the common understanding that it is about the controlling of scope changes. The real definition has two main parts: (1) the prevention of ***unnecessary*** *changes and (2) maintaining traceability of all approved changes.*

PROJECT TIME MANAGEMENT

An important item to remember is that t organizational process assets and enterprise environmental factors are inputs to many processes in each of the *PMBOK® Guide* knowledge areas. These items appear several times within the Project Time Management sub processes. Project Time Management includes: activity definition, activity sequencing, activity resource estimating, activity duration estimating, schedule development, and schedule control. The 4th Edition lists them as define activities, sequence activities, estimate activity resources, estimate activity durations, develop schedule, control schedule. The general principles have not changed. These processes are not necessarily listed to be performed in sequence, although there is some logic to how they have been arranged in the *PMBOK® Guide*.

Activity Definition

The major inputs to activity definition are the, scope baseline, enterprise environmental factors and organizational process assets. The 3rd Edition included the project scope statement, the WBS, WBS dictionary, and the project management plan. The scope baseline includes all of these items.

Keep in mind that inputs to one process can also be inputs to other processes in the same or other knowledge areas.

Activity Definition Tool and Techniques

The process of decomposition, as previously explained breaks the WBS down further into smaller components and eventually into work packages. A work package is the lowest level deliverable in a branch of the WBS and is comprised of activities that will be assigned to performers. These activities will be placed in a logic sequence that will be used to develop the project schedule.

Rolling wave planning is also listed under tools and techniques and is a form of progressive elaboration where the near term work is planned at the work package level and later work is planned at a higher task level. Planning components—these are important elements of a project plan that require more detail and should be included in the plan basically as a place holder. As more information is obtained the planning component is further elaborated and through the decomposition process, the work packages are defined.

Activity Sequencing or Sequence Activities

There are three basic types of dependencies: mandatory, also known as hard logic (due to physical requirements or limitations it must be done in this sequence); discretionary or soft logic (generally associated with best practices), and external factors outside of the project including other projects, government regulations and non project activities that may impact your project. Activity sequencing is a major part of the project scheduling process and produces the network diagram. The network diagram shows predecessor/successor relationships, logical relationships and the flow of the project work.

The precedence diagramming method (PDM) creates a network diagram of the project and displays the logical relationships of each activity. The four logic relationships in PDM are:

1. Finish to start—this is the most common form of logic relationship. Most project activities will be displayed as FS (finish to start) meaning the previous activity must be completed before the next activity in the sequence may start.
2. Start to start—This relationship indicates that two or more activities are related and may start at the same time. In some cases the plan requires some lag or delay before the start of the related activity.
3. Finish to finish—This relationship indicates that two activities may complete at the same time. There may be special considerations that drive this logical relationship. Again, some lag or delay may be involved.
4. Start to finish—A rare type of logical relationship in which the completion of an activity depends on the start of a predecessor. An example might be: The completion of a training program (finish part of the relationship) depends

on the start of a system installation. In other words, training can not be completed until after the system has begun to be installed. There may be some lag associated with this relationship also.

Activity Resource Estimating or Estimate Activity Resources

To effectively estimate activity resources the project manager and team may use their expert judgment, conduct an alternatives analysis (determining if there are other options to consider), review published estimating data (information that is available from previous projects, PMO records, or commercial data bases), and bottom-up estimating. The bottom-up estimating technique utilizes the WBS as a means of developing reliable estimates. The work packages are defined to the desired level of detail through the decomposition process. This level of detail enables the person or group responsible for submitting the estimates to develop a realistic estimate of resource requirements. Generally the work packages are defined to a level of detail where they require 40 or 80 hours of effort to complete. Reminder—many of the planning processes are done concurrently or overlap to some extent. The processes described in the *PMBOK® Guide* are not necessarily meant to be performed sequentially. Developing the WBS, estimating resource requirements and developing duration estimates may be accomplished in one planning session.

It is also important to consider the resource calendar to determine availability. Many organizations maintain a "resource pool" that provides project managers with a view of the organization's resources, skill levels and availability. In the International environment, the resource calendar becomes an even greater planning factor.

Activity Duration Estimating

To **estimate** activity duration, it is important to determine resource requirements first and to determine the skills and capability required of each resource. Remember to consider the availability of the resources, the type of organizational structure, and any historical estimating information (lessons learned) when developing estimates. In a matrix organizational structure there is almost a certainty that the resources will be interrupted from completing their assigned work due to other priorities. The inability to complete work in a continuous manner generally leads to additional time required or something commonly referred to as ramp-up and ramp-down time. In the early stages of a project life cycle an analogous estimate may be developed using previous, similar, projects as a basis for the estimate. Analogous estimates are referred to as top down estimates and should not be considered to be reliable or realistic. Analogous estimates should be considered as a starting point only. It is extremely rare that two projects would be identical. A parametric estimate using historical data and quantifiable information may be used for some projects. As an example this type of estimate may use the number of labor hours per unit of work multiplied by the rate for the labor (12 hours x 4 resources at $100 per hour). Other factors to consider when developing activity duration estimates include, reserve time (contingency for possible risks and uncertainty), calendars (local and worldwide) for availability, resource skill levels, and the general attitude of the work force. The use of mathematical equations may also be used. Three point estimating using optimistic, most likely, and pessimistic estimates provided by the functional managers assigned to the activities can provide estimates that have some degree of reliability. The use of a weighted average estimate where the most likely time is given a weighting of 4 times its value may be applicable for use on some projects.

The formulas for three point estimating are expressed as:

a) Triangular distribution: Optimistic time + Most likely time + Pessimistic time. The sum is divided by 3. This will result in the average time for completion.

$$\frac{a + ml + b}{3}$$

b) Beta Distribution: Optimistic time + 4 times the most likely time + pessimistic time. The sum is then divided by 6 to produce a weighted average time

$$\frac{a + 4(ml) + b}{6}$$

where a = optimistic, b = pessimistic and ml = most likely

Schedule Development or Develop Schedule

When developing the actual project schedule, you must consider availability of resources, capability of the resources, number of resources required, the work calendar of the resources (consider international issues such as holidays and local customs), skill levels, vacation time, possible sick time, external issues (such as weather), lead or lag requirements, and the need for resource leveling (balancing work among resources when an over allocation of work exists. Schedule compression techniques may also be applied to meet predetermined project end dates or changes in the expectations of stakeholders. The two most common forms of schedule compression are fast tracking (the overlapping of work normally done in a series or in a mandatory logical relationship) and crashing (adding resources to an activity to reduce the activity duration).

During the schedule development process the critical path method may be utilized. In this process a network diagram is created through a team effort to display the logical relationship of all activities. The estimated activity durations are applied to the network. Project software will calculate the longest path through the network. This path, known as the critical path, will have no flexibility regarding changes in activity duration, Slipping a completion date of an activity on the critical path will cause the project end date to slip. The process of performing a forward pass and a backward pass through the network will indicate the early start and finish and latest start and finish for each activity. This will identify where float (or slack) exists. Float provides some flexibility in completing activities that are not on the critical path and could be used to re-plan the network and move resources to protect critical activities.

It is also helpful, during schedule development, to conduct "what-if" scenarios (an exercise designed to create possible situations that could affect the activity of the project in general) to assist in managing potential risk events and to plan contingencies.

Critical chain or theory of constraints is another technique that may be used for schedule development. This method focuses on the resource constraints associated with the project. Identifying the weakest links (the resource issues), then strengthening the weak areas and managing through-put (the work processes that produce results) will, in theory, reduce overall project duration.

Schedule development produces the schedule baseline—this is the approved project schedule and becomes part of the project management plan.

Schedule Control or Control Schedule

Schedule control, a subset of integrated change control, involves reviewing project performance, analyzing schedule variances, and determining if corrective action is required.

Earned value management is a common technique used to assess the project in terms of schedule and cost performance to identify variance from the baseline. If the schedule variance is considered to be unacceptable, the project manager and team may use schedule compression, what – if scenarios, and resource leveling to return the project to acceptable performance levels. Some variance from the plan can be expected. An acceptable level of variance is usually established during the planning process and is based on the type of estimating techniques practiced by the project team.

Terms and Acronyms Used with a Network Diagram

At the completion of the "sequence activities" process a network diagram is produced. The network diagram displays the logical relationships of all project activities and become and input the "develop schedule" process. There are a number of terms and acronyms associated with the network diagram.

- **PDM**. The precedence diagramming method has four logical relationships: FS–finish to start, FF–finish to finish, SF–start to finish, and SS–start to start. The FS relationship is the most common relationship in the project network.
- **AOA**. Activity on the arrow. This type of network diagram is rarely used today, the PDM method is prevalent in today's project software. The AOA method uses only a finish to start relationship and may use *dummy activities*. The dummy activity indicates that a logical relationship exists between two activities but does not consume resources or have duration. It may be a factor in determining the critical path.

- **CPM**. Critical path method: Determines early or late start and finish times by completing a forward and backward pass through the network. The process will determine the critical path or longest path through the project network.
- **Slack**. Also known as *float*. It basically means flexibility. This is the amount of time an activity can slip without affecting the project end date. There is generally no float or slack on the critical path.
- **Critical path**. This is the longest path through the network that determines the earliest completion date of the project. The critical path is determined by performing a forward pass and a backward pass through the project network. Generally speaking, there is no float or slack on the critical path and any slippage of an activity on the critical path with affect the project end date.
- **Forward pass**. Determines early start and finish dates. The forward pass starts at the beginning of the project and moves through each activity (left to right in the network diagram, identifying the early start and early finish of each activity). Where there are converging points (several activities converging on one activity), the early start of the activity at which the other activities converge is determined by the greatest value of the early finish calculations of the converging activities (use the larger number at points of convergence).
- **Backward pass**. This process determines the late start and finish dates. The backward pass begins at the end of the project and moves backward (right to left) through the network, identifying the latest finish and latest start of each activity. Use the lowest number of points of convergence during the backward pass process.

- **PERT** — Program evaluation review technique. Uses a weighted average formula: (Optimistic + 4 Most Likely + Pessimistic)/6. This formula is used to determine the range of an estimate and a likely outcome based on the input of the functional managers who provide the values for use with the formula (use of expert judgment and previous experience).

PROJECT COST MANAGEMENT

Cost management involves the development of estimates, in terms of dollars or currency, for the work to be completed and for the resources that will be used on the project. The resources include equipment, materials, supplies and the people who will perform the work. There are three main processes for cost management:

1. **Cost estimating (Estimate Costs as shown in the 4th Edition).** This is the process used to determine the approximate costs of all project resources and project work. Project costs may also include travel, rent, leases, and support services.
2. **Cost budgeting (Determine budget).** Aggregating the estimated project costs to develop the project cost baseline. In this process the estimated costs for each work package and element of the WBS are summed up to produce the project budget. The costs are then allocated across the project life cycle and assigned to each phase as needed to support the scheduled work. This creates a time phased budget.
3. **Cost control (Control costs).** This process is used to monitor project cost performance and manage variances and changes to the cost baseline.

The *PMBOK® Guide* includes an overview chart for each knowledge area. These overview charts provide an "at a glance" view of the processes associated with each knowledge area. Review and become familiar with inputs, tools and techniques, and outputs of each process. Remember, as you review the *PMBOK® Guide* you will notice that many of the inputs, tools and techniques, and outputs are used in processes found in several chapters and in sub processes within a specific knowledge area. This further indicates the importance of an integrated or systems approach to managing projects.

The cost management processes will produce the cost management baseline. During the planning process the desired precision level of cost estimates (example: rounded off to the nearest dollar) will be determined, the units of measure (type of currency as an example) that will be used will be agreed upon and the links to organizational procedures will be defined (accounting and financial processes, established reporting and control procedures, industry standards).

Cost Estimating or Estimate Costs

Effective cost estimating requires the appropriate inputs such as the scope baseline, the project schedule, the human resources to be engaged and a risk register. A risk register is an organized and prioritized list of potential risk events. Estimating the cost of the project may involve the use people from various functional entities (functional managers) who provide expert judgment through experience. Analogous, parametric, and bottom up estimating may also considered.

- Analogous estimating – comparing the current project to previously completed similar projects (a top down estimating technique).

- Parametric estimating—a technique that uses a statistical relationship between historical data and other variables (square footage, lines of code, to calculate an estimate for activity parameters such as scope, cost, and duration.
- Bottom-up estimating—the WBS is used to create a desired level of detail. The costs are then summed up from the work packages to develop a project cost estimate. This type of estimating is also known as grass roots or engineering estimates.
- Incremental estimates—high level or rough estimates are produced in the early stages of the project. As planning details emerge task based estimates are produced creating more reliable estimates. This may also be referred to as progressive elaboration.

It is important to consider the accuracy of the estimating technique, the level of effort required to produce the estimate and the phase in which the estimate is being developed. Analogous estimates are considered to be high level and not accurate but can provide a "ball park" or rough order of magnitude type estimate. Parametric estimates are generally considered top down but can produce a high level of accuracy depending on the quality and sophistication of the data built into the model. Bottom-up estimates produce a more "definitive" estimate depending on the level of detail developed in the WBS. It is always a good idea to consider the risks, advantages and disadvantages of each technique when engaged in the project cost estimating process.

Estimating tools and techniques such as analogous, parametric, and bottom-up estimating will produce various accuracy levels in cost estimates. A general rule most project managers follow is to engage people who actually do the work to develop cost estimates.

In addition to analogous, parametric, and bottom up estimating there are a number of other tools and techniques that may be utilized in the cost estimating process. These include:

- *Vendor bid analysis.* Cost estimates provided by contractors may be verified through vendor bid analysis. Reviewing and analyzing estimates received from vendors and suppliers will assist in determining a more reliable overall project cost estimate. This technique will help to determine if an estimate is fair and reasonable when compared with previous projects or other bids.
- *Cost of quality.* Consider the investment in quality when determining estimates. The cost of quality is divided into two main parts:
    1. Cost of conformance—preventing defects, training, and inspections. These may all increase the cost of the project depending on the level of quality desired in the product.
    2. Cost of non-conformance—repairs, defects, rework and scrap. These items may result in greater project costs through recovery efforts and poor quality outputs.
- *Reserve analysis.* Reserve analysis is a process used to analyze the contingencies that have been added to an estimate. Adding a schedule contingency reserve or adding additional funding to a task to manage possible risk events will increase overall project cost. Reserve analysis is used to ensure that the appropriate reserve has been added.
- *Life cycle costing.* This process includes the total costs associated with the product from conception to retirement of the product. The project life cycle is included within the total life cycle costs. Project managers are generally

concerned with the project life cycle costs. Maintenance costs and other costs incurred after project completion and hand-off considered part of the total life cycle costs of the delivered product.

Cost Budgeting or Determine Budget

In this process, the project manager applies or allocates cost estimates across project phases to create a cost baseline. This is creates a time phased budget and is used to monitor project cost performance.

During the process of determining the budget the project manager may be required to address the fact that there may be a funding limitation applied to the project. If the project cost estimate is greater than the funding planned by the sponsor or executive decision makers the project manager may become involved in a process known as funding limit reconciliation. In this situation the project managers negotiates with key stakeholders to attain a solution.

The **cost baseline** is the approved project budget and is used to monitor project cost performance by comparing actual cost to the baseline and determining where variances may be developing.

Cost Control

During project execution the project manager and team will monitor project performance examining variances and ensuring that results are within planned parameters. Some changes in plans can also be expected during the project life cycle. Managing change requests is done through a cost change control system. Cost control includes the use of performance data to identify variances and may also include project reviews, variance management, and project performance reviews. Earned value management is commonly used to identify variances and positive or negative trends in project performance and to determine appropriate corrective actions.

Cost estimate updates are an output of cost control. You will notice that updates to many items (such as organizational process assets updates) will be found in the outputs section of each major knowledge area process group. This is just a reminder that a continuous learning process should be in place and new lessons learned may be discovered as each process is executed and managed. These lessons learned can provide the organization with useful changes to existing procedures and should be communicated regularly.

Key Terms to Remember

- **Work performance information—**Progress of the project including completed deliverables, deliverables in progress, incurred costs.
- **Variance analysis**
- **Earned value measurement—**a Performance measurement method that integrates scope, schedule, and cost. There are three key dimensions involved in earned value measurement—Planned Value, Earned Value, and Actual Cost.
- **Forecasting—**Determining the probable outcome of the project in terms of budget. An analysis of current performance and trends may produce an Estimate and Completion (EAC) which is compared to the original budget to determine a project cost variance.
- **Learning curves** reflect improvement with repetition. The basic premise of the learning curve is that

productivity improves through repetition. The rate of improvement decreases during the process as productivity is measured. Each time an activity is completed, there will be an improvement in productivity, but the rate of productivity will decrease as the activity is repeated. Eventually, the rate of improvement will level off, but these data can be used to estimate overall production output.

Cost Management Formulas and Related Terms

Project cost management is associated with the development of a project budget but also includes the techniques utilized to select projects or make decisions about whether or not a project should continue into the next phase. The following formulas may be used in the decision processes for selection or approval to continue:

Future value, $FV = PV\ (1 + r)^n$—Used to determine the future value of an investment. FV = Future value, PV = Present Value, R = Interest rate, n = number of years

Present value, $PV = FV\ /\ (1 + r)^n$—Used to determine the present value of a future investment. This formula discounts the value of the investment to reflect present dollar value

Net Present Value, NPV = the sum of all terms $Rt\ (1 + i)^t$

Where t = time of the cash flow, I = discount rate

R_t = net cash flow (inflow – outflow)

NPV is the sum of all present values minus the initial investment. The formula is commonly expressed as:

$$\mathrm{NPV} = \sum_{t=1}^{n} \left[\frac{\mathrm{FV}_t}{(1+r)^t} \right] - \mathrm{II}$$

If NPV is greater than or equal to zero, the project is generally considered acceptable. A NPV of greater than zero is desirable. To determine the NPV, you must calculate the present value of cash flows for each year of the project. The present values are summed up and then the initial investment is subtracted from the sum. This will indicate if the project is actually going to produce revenue.

IRR is the **internal rate of return.** This is an iterative process to determine the actual rate of return of an investment. The IRR is the rate at which the NPV becomes zero. The IRR generally must meet or exceed a predetermined "hurdle rate" established as part of the organization's project selection criteria. It requires additional effort but will provide more accurate data about the projects being considered for investment.

Payback period is the time period required to recover the initial investment. This method does not generally consider the time value of money.

Break-even analysis is the point at which cash outflow and in flow are equal.

Sunk costs are money already spent. Sunk costs are not considered when making decisions about continuing to fund a project or to move into the next phase of a project.

Project managers have to consider **indirect costs** versus **direct costs**—Labor and material are considered direct costs to the project. Indirect costs may be associated with the costs of maintaining a building, such as rent, electric power, and office support. Indirect costs are those costs that are incurred for common or joint objectives and therefore cannot be identified readily and specifically with a particular project.

Opportunity Cost—an economic opportunity that is lost as a result of a decision by an organization. Generally an opportunity cost is associated with the loss of possible revenue in one area when funding is allocated to another area or held back for other business reasons.

Earned Value Formulas and Related Terms

Earned value analysis or earned value management provides the project manager and team with a "point in time" view of a projects performance. Earned value analysis allow the project manager to calculate project variances by comparing planned performance against actual performance. There are three main factors to consider in earned value analysis:

1. *EV* or earned value also expressed as *BCWP* or budgeted cost of work performed – This is the estimated cost of the work that has been ***completed***.
2. *PV* or Planned Value also expressed as *BCWS* or budgeted cost of work scheduled – This is the estimated cost of the work that ***should have been done***, according to the plan, at the time of measurement.
3. *AC* or actual cost, also expressed as *ACWP* or actual cost of work performed – This is the actual amount paid for the work that has been completed at the time of measurement.

The commonly used earned value formulas include:

EV or $BCWP - AC\ (ACWP) =$ Cost variance or, CV

A negative cost variance generally considered to be unfavorable and undesirable. Example: $EV = 1000, AC = 1200$. The cost variance is -200.

$EV\ (BCWP) - PV\ (BCWS) =$ Schedule variance, SV

A negative schedule variance indicates the project is behind schedule: Example" $EV = 1000$, $PV = 1200$, The schedule variance is -200 or 200 units of measurement behind schedule

EV or $BCWP\ /\ AC =$ Cost performance index or CPI

This formula will provide an indication of the efficiency in the management of project costs. A cost performance index equal to of 1 indicates that the project has no variance in cost and is being managed to the plan. Example $EV = 500$ and $AC = 500$. $500/500 = 1$

EV or $CWP\,/\,BCWS$ = Schedule performance index, SPI

This is an indication of efficiency in managing schedule performance. A schedule performance index (SPI) of 1 indicates that there is no schedule variance and the project schedule is being managed effectively.

A performance index equal to 1 means the project is on plan with no variance. An index greater than 1 means the project is performing better than the plan; an index less than 1 means that the project is not performing as well as had been planned. This information must be interpreted in the context of the plan to determine where variances may be acceptable and what the variance tolerances are for the project and or the organization.

PROJECT QUALITY MANAGEMENT

The main objective of quality management is customer satisfaction. Quality, for the purposes of defined project success, is always defined by the customer. The project manager and the team may be required to work very closely with the customer to define the specific needs of the customer and to develop and then refine project requirements. There are three major processes in quality management: Quality Planning or Plan Quality as expressed in the *PMBOK® Guide* 4th Edition, Perform Quality Assurance, Perform Quality Control.

Quality Planning or Plan Quality

Quality planning is focused mainly on achieving customer satisfaction. In addition the project manager and team should attempt to manage quality by preventing errors from affecting the project deliverables, instill a working environment of continuous improvement, and ensure that responsibility for quality is delegated to the appropriate individuals. It is the responsibility of the supporting management team to ensure that the project team and project performers have the necessary training, tools, and equipment to meet quality objectives and achieve project success.

Project managers and team members must maintain an awareness of the cost of quality. The cost of quality includes the cost of conformance (prevention and appraisal) and the cost of nonconformance (internal failure, or rework, and external failure, or customer dissatisfaction). It is less costly to prevent errors than to inspect and identify errors that must be corrected. Thus, organizations need to have a *policy* of quality—higher-level guiding principles of quality within an organization. A project quality policy may be developed based on the *organization's* overarching quality policy.

The quality planning process produces a quality management plan, quality metrics, quality checklists, a process improvement plan, a quality baseline, and updates to the project management plan. Inputs to quality planning include the scope statement and project management plan. Remember to consider organization process assets and enterprise environmental factors, as well. A quality plan is a subsidiary of the project management plan.

Project Quality management includes but is not limited to the following:

- **Continuous improvement.** Quality is about continuous improvement. The Plan, Do, Check, Act cycle known as the Shewhart Chart forms the basis for the five major project management processes.
- **The contributions of quality experts**:
  - Deming—Statistical analysis to encourage continuous improvement and management's responsibility to support

the workforce and provide the necessary tools and training to ensure high levels of quality.
    - Juran—Fitness for use and safety of products.
    - Crosby—The pursuit of zero defects and avoiding the cost of non conformance.
- **ISO 9000/2000.** The International Organization of Standardization sets international standards that often become law. Establishing and then adhering to processes approved by IOS will produce a consistent output.
- **Six Sigma.** First introduced at Motorola in 1986, Six Sigma sets a standard of allowing only 3.4 defects per million opportunities for failure. The methodology for meeting those standards is to follow a procedure refereed to as DMAIC: define, measure, analyze, improve, control.
- **TQM, Total quality management.** TQM is an organizational approach to quality that starts at the top management level and includes all levels of employees. The focus is on continuous improvement and everyone is responsible for quality.

It is management's responsibility to provide the resources for effective quality planning. The right tools, equipment, and training are needed to ensure the appropriate level of quality for the project deliverable is attained. The following are tools and techniques that may be used in quality planning:

- **Cost / benefit analysis.** Determine the level of quality to be included in the product or service and return on the investment. This is a ratio of the costs of the improvement to its expected benefits.
- **Benchmarking.** This process uses the best in class as the standard against which to measure existing

performance. Use benchmarking to identify gaps and to target areas for improvement.

- **Flowcharting.** Diagramming a process can define the steps in the process and identify gaps and redundant work. Flowcharts provide an understanding the overall operation or flow of work.
- **Design of experiments.** This is generally associated with GenichiTaguchi. Quality should be built or designed into a product, not inspected into it. The further away from the desired target (level of performance or quality) the more expensive it will be to return to the target area.
- **Control Charts.** used to determine if a process is stable by plotting measurements and comparing with a determined mean value and pre-determined upper and lower control limits.
- **Cost of Quality.** The costs incurred over the life of the product associated with correcting non conformance issues as well as well prevention and appraisal costs.

Quality Assurance

Quality assurance is the implementation of systematic activities to ensure that the project will satisfy quality standards. Quality assurance tools includes:

- Audits—independent reviews of project activities and related work to determine if organizational policies and standards are being observed and utilized appropriately.
- Process analysis—examination of process steps to identify gaps and areas for improvement.

Quality Control

Quality control involves monitoring project results to determine if they comply with quality standards. The focus is on prevention—keeping errors from the customer.

The PMBOK® Guide 3rd and 4th Editions list ten common quality control tools: page 169 of the 4th Edition draft and page 192 of the 3rd Edition.

1. **Inspections.** Inspections are a key tool for verifying an output. Inspections may also include walk-throughs of an area, testing, and periodic reviews.
2. **Statistical Sampling.** Selecting a specific part of a product population and submitting that part for testing. There are two common forms of sampling: Attribute sampling which is a pass or fail approach. There is no flexibility, the product either conforms to specifications or it does not. Variable sampling allows for some variance from specifications and products or outputs are graded on a scale to determine acceptance. In addition, the sampling process may produce the following results:

   Producer's risk or Alpha risk—the risk to the producer that a good lot may be rejected.

   Consumer's risk or Beta risk—the risk that the consumer or buyer may purchase a bad lot.
3. **Control charts.** Control charts are used to determine if a process is in control or out of control. Data points are observed to determine if they meet requirements and are with upper and lower control limits. The rule of 7/21 is commonly used in interpreting control chart data. The rule of 7 states that it takes 7 consecutive data points to indicate a trend or a run. An example of a process out of control would be seven data points in a row above the mean. The rule of 21 states that 21 data points are needed for

the sampling to be statistically valid. These are basic rules and easy to remember but may not be acceptable criteria for many organizations.

4. **Pareto diagrams (80/20 rule).** This tool is based on the principle, as described by Pareto, that 20 percent of the issues cause 80 percent of the problems. By charting the frequency of quality issues, the most problematic quality issues can be identified and addressed.
5. **Cause and Effect Diagrams (fishbone diagrams).** These diagrams help project teams define specific problems and then provide a means to identify several potential causes of the problem. The major causes are prioritized and corrective action is determined. The diagram resembles the skeleton of a fish and was developed by Kaoru Ishikawa in the 1960's.
6. **Flowcharts.** Using symbols to represent actions or decisions, a flowchart walks the process through stages. This can help identify weaknesses in the process by showing how parts of the system interrelate.
7. **Histograms.** These bar charts show distribution frequencies.
8. **Run charts.** These are line graphs that show trends or variations over time.
9. **Scatter diagrams.** These are charts that plot the relationship between two variables. The closer the points on the diagram cluster along a straight line, the more direct the relationship is between the two variables.
10. **Defect repair review.** This is an action taken to make sure quality defects are fixed and brought into compliance with the standards.

It should become a matter of standard practice for project managers to document the lessons learned at the completion of the quality control process. This practice will benefit the project team as well as the organization supporting the project. Lessons learned are considered to be part of a project manager's professional responsibility.

HUMAN RESOURCE MANAGEMENT

The human resource (HR) management knowledge area includes what is considered to be the "soft side" of project management. This includes organizational structure, planning for human resources, acquiring the project team, developing the team, and managing the team. Human resource management is associated with leadership, motivation, team building, managing conflict, establishing relationships with functional groups, communicating effectively, and providing useful feedback regarding project performance and individual performance. For many project managers, the skills necessary to manage the project team present the greatest challenges. Dealing with the human side of project management and the emotional issues requires patience, good listening skills, good judgment and strong leadership.

Human resource management includes an understanding of organizational influences and the difference between project-based and non–project-based organizations. The project-based organization generally provides goods and services to clients through a project management methodology. The non–project-based organization does not have a project management methodology in place and may not require a formalized project approach due to the nature of the business.

Organizational structure

There are three main types of organizational structure:

1. Functional (also known as stove pipe, chimney, and silo). A manager is assigned responsibility for employees with a specific set of skills or expertise.

2. Matrix (there are three types of matrix environments—weak, balanced, and strong). In the weak matrix the project manager has very little authority and must rely on interpersonal skills and influence to achieve objectives. The balanced matrix may create a competing environment where the project managers and the functional managers have equal levels of authority. There is a potential to be exposed to the **two-boss syndrome** where resources may be receiving direction from the project manager and the functional manager. This could cause some conflicts regarding priorities and the assignment of resources.
3. Projectized. In this form, projects are separated and have distinct teams that report directly to an assigned project manager. In the projectized structure, the project manager has a significant level of authority. The project stakeholders are generally defined as anyone involved in or directly impacted (positively or negatively) as a result of the project. Key stakeholders include the project manager, sponsor, customer, user, and project team.

Types of Power

There are five general types of power: Formal or legitimate (given in connection with a person's position in the organization), reward power—the power of influence one may posses by having the ability to offer items of value to influence another person, penalty power—the ability to discipline or in other ways create an uncomfortable environment if the employee does not follow directions, expert power (associated with knowledge and experience), and referent power. **Referent power** is individual power based on a high level of identification with, admiration of, or respect and connection to the actual power holder. This is sometimes translated into "power by association."

Tools and Techniques in HR Management

Project managers will experience conflict during the life of a project. It is important to develop an understanding of types of conflict and how to manage conflict effectively. Conflict in the project environment may involve grievances, schedule disagreements, funding requirements, resource availability, quality standards and many other issues. Project managers will be required to manage conflict effectively and develop conflict resolution skills. Conflict resolution may include mediation, arbitration, negotiation or litigation.

- **Conflict management.** There are five basic approaches to managing conflict commonly found in project management literature:
    1. **Withdrawal,** or stepping away from and avoiding the issue
    2. **Smoothing,** an attempt to minimize the severity of the conflict. This approach can be described as "playing down" the differences while focusing on common interests.
    3. **Compromise,** a win–lose, lose–win approach where each side must give up something of value to reach an agreement.
    4. **Forcing,** or using position power to solve a problem, usually results in later conflict that can be more severe. This is known as a win/lose approach.
    5. **Collaboration,** which is a win–win strategy of problem solving by confronting the issue with a view that will benefit both sides. This form of conflict handing is often associated with confrontation

where the parties involved "confront" the problem together.

Other tools and techniques associated with Human Resource Management include:

- **Responsibility assignment matrix (RAM).** This tool links project team members with WBS tasks. The RAM is used to define clear responsibility for the completion of project tasks. It is not a tool for determining resource requirements.
- **Project interfaces.** There are several types of project interfaces: organizational (business unit), technical (design group and production group), and interpersonal (formal and information personal relationships). The project manager must be aware of the interfaces and establish effective relationships with each. Most project managers depend on other organizations to fulfill resource requirements, and it is essential for the project manager to create an environment of mutual trust and respect. This will help to ensure a smooth implementation and to facilitate the process of acquiring resources.
- **Staffing the project.** The use of a staffing pool or resource pool will assist in staffing a project team. The goal is to obtain the best resources available. In most cases, the project team is composed of resources commonly classified as "competent" performers who will require some coaching, training, and mentoring from the project manager. The project manager can develop a high-performing team from an initial group of competent performers with the appropriate balance of effective management and leadership.

- **Negotiating.** Project managers must develop effective negotiating skills to achieve project objectives. Items that may require negotiation include obtaining resources, schedule development, activity duration estimating, and obtaining project funding. Negotiating involves knowing your objectives, preparing for discussions, willingness to listen, and an appreciation for the other party's position.
- **Team development.** The project manager is a team leader and should continuously use existing techniques as well as develop new and effective methods to enhance the performance of the team. Teambuilding activities such as reward and recognition events, offsite meetings, motivation building sessions, team training events, feedback sessions, performance appraisals, co-location of the team whenever possible, and joint activities that encourage support among team members can improve a project's probability of success.

Stages of Team Conflict and Team Development

There are four general phases in the team development process:

1. **Forming.** In this phase of bringing the team together, the team members are cautious and unsure of their roles. The project manager plays a dominant role here and provides direction to the forming team. Roles and responsibilities of the team members are not clearly defined.
2. **Storming.** In this phase, there is great focus on their individual needs and to be protective of their own interests. There may be some "jockeying for position and a high degree of conflict. The project manager provides

direction and support and acts as a facilitator in this phase.

3. **Norming.** In this phase, individual assignments have been defined and team members focus on their specific tasks.
4. **Performing.** In this phase, the project team is working as a collaborative and supportive team, with a focus on the higher level set of project objectives and a shared vision of success.

Motivation

Project managers face a significant challenge. In many organizations they must compete for resources, manage teams they have no direct authority over, and work with other internal or external entities that may have a different set of priorities. The project manager must become motivated, sustain that motivation, and motivate others to achieve the objectives of the project. To accomplish this, the project manager must develop and continuously enhance his or her leadership skills.

Leadership involves developing a vision, mission, and objectives and then creating a motivated team that will achieve the vision and objectives of the project. The project manager is, in most cases, considered to be the leader of the project team, whether or not he or she possesses actual legitimate power. So how does the manager go about motivating the team? An understanding of motivational theory is a good starting point.

Abraham Maslow might be considered the father of motivational theory. **Maslow's hierarchy of needs** is probably the most well known of the motivational theories. The basic premise is that as each motivating factor is achieved, it is no longer a motivator, and the person reaches higher up in a hierarchy of needs. The levels of the hierarchy include:

physiological needs, safety and security, social needs, self-esteem, self-actualization.

Douglas McGregor thought motivation could be associated with managerial styles. He defined two different styles he called Theory X and Y. In **Theory X**, managers distrust employees and believes employees do not want to work and will do only what is minimally required. A high degree of supervision is associated with Theory X. In **Theory Y,** managers trust the employees. This is a participative style of management and is more trusting and supportive. This type of manager believes people want to work and contribute to the well being of the project and the organization.

According to Frederick Herzberg, it is important to identify and remove items that causes dissatisfaction. These items are referred to as—hygiene factors and include compensation, level of supervision, the work environment and peer relationships. Problems in these areas may cause a severe loss of motivation among employees. Correcting these problem areas may not create a motivating environment. Once the problem areas have been resolved the focus should shift to motivating factors such as greater challenges, more responsibility, advancement, reward and recognition. According to Herzberg, it is important to remove the "pain points" from the environment before you can motivate the work force.

Leadership and motivation are key factors in project management and can have a major effect on a project's outcome. Project managers step into a leadership position when they accept a project assignment and should dedicate some of their time to the enhancement of leadership skills.

Leadership skills include but are not limited to the following items:

- Problem solving
- Facilitating the integration of new team members into the team

- Managing interpersonal conflict
- Facilitating group decisions
- Communicating effectively
- Presentation skills
- Creativity and innovation
- Ability to develop and articulate a vision
- Motivation and influencing skills

PROJECT COMMUNICATIONS MANAGEMENT

Project communications is about effective communication between the project manager and stakeholders. This means developing a plan and the processes to ensure that the right information is provided to the right people at the right time. Projects stakeholders may require different types of communication and in varying degrees of detail. In addition, today's project environment requires an understanding of cultural differences, technological factors, and the ability to communicate effectively with dispersed or virtual teams.

There are many forms of communication including internal organizational communications, project communications, communication with the media, formal reports, informal unstructured communication, written, verbal and non verbal communications. The project manager's role demands strong communication skills and the ability to interface with stakeholders at every level within an organization and externally.

The elements of communications management are integrated across all project management knowledge areas and with a strong set of communications skills the project manager will find it difficult to accomplish any project objectives.

In the project environment project managers communicate primarily in a horizontal or cross-organizational mode.

Transferring and receiving information from the functional groups and business entities involved in the project. The project manager position also requires communication with management (known as upward communication) and to subordinates or employees (downward communications).

Project communications is successfully achieved through an understanding of the **sender–receiver model**. In the model, the critical elements are the region of experience of the sender and receiver. This region of experience defines the knowledge, credibility, confidence, and overall experience of the sender and the receiver. (An overlap of the regions of experience where there is a shared knowledge or familiarity about a subject between sender and receiver or where each has developed similar backgrounds will facilitate the communications process.) The sender prepares or encodes a message to transmit to a receiver. The message passes through the sender's personality screen, which defines the style of the sender and could have an impact on how the message is actually transmitted. This message travels across the region of experience to the receiver. The message passes through the receiver's perception screen, where the receiver may analyze the message, the credibility of the sender, and other factors about the sender. The message is then received and decoded. To ensure that the message was received as intended, the receiver will feed back the message by encoding the message, passing it through the receiver's personality screen, through to region of experience, then through the sender's perception screen. The message is decoded by the original sender to determine if the message was sent and received as intended. Because there is a potential for some distortion of the message, this process may be repeated a few times to ensure that the message was clearly communicated.

The complexity of project communications can be determined through the use of the communications channel formula.

This formula provides the project manager with a view of the potential for communications problems or breakdowns that may occur during the project life cycle., When working with a large team of people the project manager may apply this formula to assess the number of communications channels and therefore the number of possible areas for miscommunication. The formula is expressed as:

$N(N-1)/2$

N = Number of people on the project team

Example: Project team members total 10. Applying the formula the number of communications channels is $10(10-1)/2$ or 45 channels.

Increasing the number of people on a project team rapidly increases the number of channels of communication that will exist between project team members. As the number of channels increases, the ability to manage communications effectively and to minimize communications breakdowns becomes more difficult. The use of smaller core teams and sub-teams may help to reduce the potential of communications breakdown while managing a project team.

Project communications planning is an essential part of the overall integrated project plan. It includes the processes that are required to be followed by the project team to effectively generate, collect, store, analyze, and distribute project-related information. Effective communication to project stakeholders is a key item in the successful achievement of project objectives. The *PMBOK® Guide* defined four major subprocesses associated with project communications: communications planning, information distribution, performance reporting, and managing stakeholders. The 4th Edition has been revised to indicate 5 processes: Identify Stakeholders, Plan Communications,

Distribute Information, Manage Stakeholder Expectations, and Report Performance.

Identify Stakeholders

As stated previously in the Human Resource Management section, stakeholder are the organizations or individuals either involved in the project or affected by the project. Identifying each project stakeholder and determining their specific influence level and their interest in the project is an important factor to consider in the overall planning process. Stakeholders can affect decisions about requirements and change requests and each may have their own specific communications needs and expectations.

Communications Planning or Plan Communications

Two key tools in communications planning are communications requirements analysis and communications technology. **Communications requirements analysis** looks at the specific needs of the stakeholders in terms of type of information to be provided and format of the information (depending on the responsibility and involvement of the stakeholder).

Communications technology. The type of technology used by stakeholders may vary tremendously. Technology factors include how information will be transferred to stakeholders in terms of information systems, infrastructure, and the capabilities of the stakeholders to access and send information.

The major output of the communications planning process is the communications management plan, another subsidiary plan to the project management plan. When developing a communications plan, consider the needs of the stakeholders, the urgency of the information that will be shared, the availability and reliability of the technology, and complexity of the project. Consider the constraints (limitations) that may affect the ability to communicate, especially in a virtual or international project environment.

Project communication is often affected by the assumptions made by the project team. Assumptions are things that are considered true, real, or certain fro planning purposes. Any time a list of assumption is developed, the project manager and team should take the time to review and validate these assumptions to avoid communications errors, risk situations and surprises later in the project.

Information Distribution or Distribute Information

Information distribution focuses on providing the appropriate information to the stakeholders in a timely manner. Written and oral communication, listening and speaking, presentations, memos, formal meetings, and informal meetings may be utilized to distribute information. The project manager should be aware of the different requirements associated with internal and external communication and formal and informal communication.

Major inputs to information distribution:

- The project management plan which will, in most cases, include the communications management plan.
- Performance reports—information about the progress of the project and its current condition. The tools and techniques for information distribution include the project manager's communications skills, an understanding of the communications sender–receiver model, feedback loops and the ability to communicate verbally and through written messages. Tools and techniques

also include electronic systems, internal organizational networks, other forms of technology and types of media.
- An effective information distribution process will produce: Reports, lessons learned, project records, important project information such as memos, meeting minutes, and other documents.

Manage Stakeholder Expectations

The project manager should clearly establish the expectations of the key stakeholders. The expectations will vary among stakeholders but the project manager should make the effort to intentionally define and establish them and review them on a regular basis. This will help to prevent misunderstandings and disappointments later and help to maintain a good working relationship with all stakeholders. The interpersonal skills of the project manager, an understanding of organizational protocol, and some diplomacy are essential when managing stakeholder expectations.

Managing stakeholders is also associated with the area or domain known as *professional and social responsibility*. In this domain of project management, the project manager must balance the needs of the project stakeholders through an understanding of their specific influence and concern about the project. Project-related decisions should be made with consideration for the larger community of project stakeholders to minimize conflicts and manage bias.

Performance Reporting or Report Performance

Performance reporting involves collecting, reviewing, analyzing, and then disseminating the information to the appropriate stakeholders. Project status reports are generated to indicate the health of projects. Sometimes this report uses colors to differentiate project status conditions: red for a troubled project, yellow for a project that is experiencing some serious problems, and green for projects in relatively good shape.

The project management plan, work performance information and budget forecasts are major inputs when developing project information. Common tools for assessing and reporting project performance include earned value analysis or variance analysis. These tools enable the project manager and team to assess the project using mathematical formulas S-curve charts to indicate variances to the plan and assist in developing strategies to correct unacceptable variances or cost to return a project to the appropriate performance level.

RISK MANAGEMENT

Risk management is something that should be included among the activities of the project manager and the project team from the beginning of the project during initial planning through completion and closeout of the project. Effective project planning includes continuous focus on the threats, as well as the opportunities associated with the project. A proactive approach to risk management in which the project team and the project manager actively discuss potential risk situations will make the difference between a smooth-flowing project and a project filled with surprises and potential disasters.

Companies face two basic types of risk situations: insurable and business. Insurable risk is a calculable risk that can be spread over the many participants who purchase the insurance. With insurance risk, there is only a chance for loss. This would include damages to property or equipment and workman's compensation, for example. Business risk is the potential that a company's fortunes may rise or fall based on outside events such as competition or economics factors. When developing risk management plans the project manager should consider both types of risk. It is also important to view risk not just from a negative perspective that focuses on threats to the project or organization but on potential opportunities as well. Another equally important factor in risk management is the need to ensure that risk management is budgeted into the project as you would with any other activity. However, the amount of time and money spent performing risk management activities should be commensurate with the risk management budget and the needs of the project.

Project risk management includes planning for risk, identifying risks, performing qualitative analysis, performing quantitative analysis, planning risk responses, and monitoring and controlling risks. These processes are all part of the risk management plan. The sequence of these processes appears to a logical approach to managing risk. When the risk have been identified by the project team, the potential risks are assessed in terms of probability and impact to determine a risk rating. This is accomplished through qualitative analysis. This form of analysis is largely based on expert judgment. The risk events are prioritized and may be assessed in greater detail with more comprehensive tools and techniques in the process of quantitative analysis. The prioritized risks are analyzed and responses are developed. A risk monitoring and control process assists the project team in managing new risks or residual risks.

Some key elements of risk management that should be considered include the following:

- *Risk is a measure of uncertainty and is determined by assessing probability and impact.* Project managers deal mainly in an area or segment of the risk spectrum known as relative uncertainty. There are very few items that have a 0 percent chance of occurring and there are very few items that are considered to be 100 percent certain. Relative uncertainty basically means that some information is known but there is also some degree of uncertainty that requires the development of a risk management strategy to respond to the uncertainty.

- *Risk includes factors that may be categorized as known unknowns and unknown unknowns.* A risk in the known unknown category can be planned for to some extent. These are risks events that have occurred previously and there is some data available to work with. Risks in the category of unknown are those risks for which no information is available. This type of risk cannot be anticipated, because there are no data or indication that it could actually happen. You cannot plan for something that is not known at all. Many organizations establish Management reserves (a back up reserve of funding) to address these types of risks. An example would be a company that keeps an amount of money equivalent to their operating budget in reserve in case or a catastrophic event.
- *There are two primary components of risk: probability and impact.* A probability / impact matrix is used to rate risks by using previous data and expert judgment to determine the probability that an event may occur and the impact to the project if the event actually occurs. This technique may use scales to determine low, moderate, high ratings for probability and impact. Some organizations may have numerical values from probability and impact. This process will assist in determining which risk events have the greatest priority in terms of their urgency and potential effect on the project.

- Risk Tolerance the organizational culture may have an effec*t on risk tolerance.* The risk tolerance of an organization or an individual may be indicated through a utility factor. The utility factor associated with risk tolerance is generally defined as follows: Utility, level of comfort or the satisfaction level associated with risk rises at a decreasing rate for the risk-averse person. This means that as risk increases, the risk-averse person becomes more and more uncomfortable with the situation. The utility factor or level of comfort associated with risk increases for the risk seeker when greater levels of risk are experienced. Risk seekers thrive on risk opportunities and feel very comfortable in risk situations. Simply put, they like taking risks. Consider the culture of an organization as it relates to risk management. A risk adverse executive team will influence the employees of the organization to minimize their willingness to take risks. This is actually an enterprise environmental factor to consider.
- **Hurwitz and Wald criteria.** The Hurwitz and Wald criteria (maximax and maximin) describe the risk tolerances and behaviors of an organization regarding risk. The Hurwitz, or maximax, criterion is associated with a "Go for broke" approach where there is a belief that the outcome will be very positive. It is an optimistic outlook. The Wald, or maximin, criterion is associated with people or organizations that are primarily concerned with loss. They are considered to be risk averse.

Risk Identification

Risk identification is the process of determining which risks might affect the project and documenting the risk characteristics.

During the risk identification process the project team may observe risk triggers. These are symptoms that may lead to a risk event. It is important to identify and react to triggers before a risk event occurs.

Tools associated with risk identification include the following:

- **Brainstorming.** Managers generate risk data from the project team by encouraging open discussion of ideas.
- **Delphi technique.** This is the use of subject matter experts in an anonymous setting to eliminate bias.
- **Nominal group technique.** This is basically a combination of the Delphi and brainstorming techniques. The information is gathered anonymously, but the participants are known to all.
- **SWOT analysis.** This technique involves the identification of an organization's strengths, weaknesses, opportunities, and threats. It involves specifying the objective of the business venture or project and identifying the internal and external factors that are favorable and unfavorable to achieving that objective. The technique is credited to Albert Humphrey, who led a research project at Stanford University in the 1960s and 1970s using data from Fortune 500 companies.

Tools and Techniques for Risk Management

- **Decision trees.** Decision trees are associated with quantitative risk analysis. This tool assists in determining the most appropriate solution or strategy based on available data and forecast information.
- **Expected monetary value.** EMV is the total of the weighted outcomes (payoffs) associated with a decision,

the weights reflecting the probabilities of the alternative events that produce the possible payoff. It is expressed mathematically as the product of an event's probability of occurrence and the gain or loss that will result. Calculating the EMV will help decision makers determine which approach is most beneficial to the organization.

- **Monte Carlo process.** This is a technique that creates a simulation of a project using specifically designed software. The project is simulated many times using variable data obtained from reliable sources to determine possible outcomes and probability of task completions, schedules, and costs. This is a form of quantitative risk analysis.
- **Probability distribution.** There are several types of distribution curves: normal distribution (bell-shaped curve), triangular distribution, and beta distribution, for example.
- **Standard Deviation.** This is a statistic used to measure confidence level. It is generally calculated by the formula:

$$\frac{\text{Pessimistic} - \text{Optimistic}}{6}$$

In association with the PERT Weighted Average Formula.

Risk Response Strategies

There are seven risk response strategies: avoidance, transference, mitigation, exploitation, enhancement, sharing, and acceptance:

1. **Avoidance.** The risk is too great and alternatives must be identified.
2. **Transference.** Identify a subject matter expert that can deal effectively with the potential risk and handoff the risk to that person.

3. **Mitigation.** Attempt to reduce the effect of the risk by reducing the probability or impact.
4. **Exploitation.** Create situations that will actually cause a positive result to occur.
5. **Enhancement.** Obtain the greatest benefit of the positive occurrence.
6. **Sharing.** Inform other organizational entities about the positive results.
7. **Acceptance.** The risk has been identified and will be managed by the project team if it occurs. Passive acceptance means no special action will be taken. Active acceptance means that a specific action will be developed to manage the accepted risk.

PROCUREMENT MANAGEMENT

Procurement management is a set of processes designed to obtain the resources, products, services, and materials needed to complete the project objectives. It involves developing a strategy determining what products and services can be developed internally by the organization based on capability and resource availability and what must be obtained through suppliers, contractors or vendors. The organizational process assets and enterprise environmental factors (previously discussed) will affect decisions made during the project procurement process. Generally, procurement involves a buyer and a seller. The buyer attempts to obtain material and equipment or products at the best possible price while the seller seeks to optimize a sale for profit. Negotiation skills become of critical importance when procuring goods and services and the type of contract agreed upon by the buyer and seller will be a major factor during execution of the project plan.

The procurement processes described in the *PMBOK® Guide* 3rd Edition include the following:

- Plan purchases and acquisitions
- Plan contracting
- Request seller responses
- Select seller
- Contract administration
- Contract closure

The 4th Edition of the *PMBOK® Guide* includes the following processes that are basically slight variations from the 3rd Edition.

- Plan procurements
- Conduct procurements
- Administer procurements
- Close procurements

The basic concepts of procurement management remain the same in the 3rd and 4th Editions.

As seen in other knowledge areas the sequence of these processes seems to be logical, but the actual use of these processes and the subprocesses within them depend on the type of project and the organizational policies and practices that have been defined and communicated.

The procurement planning process requires the project manager and team to consider a number of inputs. These inputs are associated with most of the other knowledge areas of the *PMBOK® Guide*. Project requirements must be identified, risk must be assessed, the type of resources required must be determined, and project scope, schedule, and cost information should be available. The project manager, if authorized, will conduct a make or buy analysis to determine what must be procured and what can be fabricated internally by the organization. There are several different types of contracts that may be considered and each contract has a set of advantages, disadvantages and risks. A knowledge of contract types is extremely important when managing the procurement process.

Types of Contracts

The key elements that form a contract are generally considered to be: an offer, an acceptance, and consideration (something of value). Contracts are legal documents and disputes and issues about the contract between the buyer and seller may be subject to remedy in a court of law.

There are many types of contracts and the terms and conditions will vary greatly depending upon the objectives of the buyer and seller. The most cases the appropriate type will be negotiated between buyer and seller in a mutually beneficial approach. Instead of opposing view points a partnership will be created that will enable each side to achieve their objectives. There are five common types of contracts associated with the project procurement process:

1. Firm fixed price (FFP) or lump sum. The contractor or seller estimates the cost of the project work and submits the bid. This type of contract can introduce a significant amount of risk to the seller and it is therefore essential to for the contractor to do a very thorough job of estimating. Expected profit and contingencies to deal with unexpected issues are included in offer made to the customer in this type of contract.
2. Cost plus fixed fee. This type of contract provides the seller with a fixed fee that is not tied to project costs. The costs, as they accumulate, are assumed by the buyer.

3. Cost plus percentage of cost. This type of contract creates an advantage for the contractor or seller. As the cost of the project increases due to changes in requirements and increase in project scope the amount paid to the seller for services increases proportionally based on the agreed upon percentage. In this arrangement there is significant risk to in the buyer.
4. Cost plus incentive fee. The buyer and seller agree to a sharing ratio where the buyer and seller both benefit if the seller manages costs effectively. The incentive is realized as a percentage of the cost savings.
5. Fixed price incentive fee. In this arrangement the project requirements are firmly set. Specific target costs, a ceiling price and a sharing ratio are negotiated. The contractor has an incentive to reduce costs whenever possible as the project is executed. A reduction in costs may lead to a greater profit. Failure to control costs or reduce costs may result in a significant loss of anticipated profit and may possibly result in further financial loss as project costs are assumed by the seller for any amount above the agreed upon ceiling price.

Other types of contracts include time and material, purchase order, letter contract, or letter of intent. The **letter of intent** provides an opportunity to move forward with the project while the details of the main contract are reviewed. This approach may help to avoid late starts, especially where critical environmental factors may impact the project (e.g., weather, location, resource availability).

Upon completion of contract negotiations a definitive contract is approved. The definitive contract includes all of the terms, conditions, and deliverables that have been negotiated between the buyer and seller.

Tools and Techniques of Procurement Management

- **Procurement documents.** Procurement documents will vary, depending on the type of project and the organizations involved. Typical documents include request for proposal (RFP), request for information (RFI), request for bid, and request for quote.
- **Statement of work.** This is a narrative description of work to be completed under contract. The statement of work in many cases precedes the project charter and the contract. It describes the work to be done and provides and opportunity for the contractor to determine if the work can be successfully performed.
- **Make or buy decisions.** These decisions are associated with whether or not the buying organization should make the required goods internally using available capacity and ability or to purchase goods and services externally. Cost, capability, and security issues are considered when making these decisions. The advantages and disadvantages of making or buying goods and services is an important part of the plan procurements process.
- **Rent vs. lease analysis.** This analysis is used to determine the break-even point in terms of cost. The project manager considers the best, most cost effective option based on the time an item is needed and the cost of each option. This is generally associated with the make or buy decision process.
- **Bidder conference.** This is an important activity in the procurement process. In the Conduct Procurements process, the bidder conference is a technique used to ensure that there is a level playing field for potential contractors.

This process includes the scheduling of a question-and-answer session about the project and provides the bidders with additional information that may be needed in their decision process.

- **Negotiation.** There are many tactics used for negotiation. In the project environment, the negotiation should be focused on a partnership arrangement where both sides can maximize the outcome to meet their needs. Location of the negotiation and the environment are critical elements. Consider the needs of each party when scheduling a negotiation meeting. The *hygiene* issues such as room size, table shape, location, and positioning of each participant are considered to be important to successful negotiation. Understanding the opposing viewpoint, using conflict management, and including problem solving strategies, is very important during negotiation. Consider the minimums that will be accepted or the maximum that will be offered when developing a strategy. Remember that fair and reasonable negotiation is expected.
- **Weighting system.** This technique is used to select a seller. Sellers are scored using various criteria, generally established by the buying organization's management, such as on-time performance, previous performance or track record, financial stability, quality, and cost. A weighting factor is used for each specific criteria. The seller is scored using the criteria, and then the weighting factor is multiplied with the score to determine a finalized score. This process provides a more balanced approach to seller selection.
- **Screening.** This process is used to ensure minimum qualifications of the seller. Failure to demonstrate the

minimum qualifications will eliminate the seller from further involvement in the selection process.

- **Special terms and conditions.** Pay special attention to penalty clauses. Failure to meet an agreed upon deliverable or achieve objectives on time may result in fines payable to the buyer. A clause for liquidated damages might say that failure to complete the project on time may result in payment to the buyer for lost revenue. A *force majeure* clause covers acts of God, strikes, and terrorism, for example.
- **Incentives.** Contracts may have provisions that will provide incentives to the buyer and the seller. An example is a fixed price plus incentive (FFPI). In these contracts, a sharing ratio is established between the buyer and seller that will provide some type of monetary incentive to keep costs down and to complete the project early.
- **Termination for convenience.** The buyer decides there is no longer a need for the product or service. There will generally be some contractual requirement for the buyer to pay for work completed, and these conditions will be expressed specifically in the terms and conditions of the contract.
- **Termination due to default.** The failure of the contractor to provide a satisfactory deliverable may result in a decision by the buyer to end the contract and seek an alternative supplier.
- **Administrative changes.** These are records only type changes and do not affect the actual work of the project. This might include a change of billing address or change of account from which funding is obtained.
- **Constructive change.** This type of change actually impacts the project work. An example would be customer-supplied equipment or property that is not available on the date scheduled. This situation usually requires replanning

and changes to the work schedule and other activities. Additional costs may result from constructive changes, and schedule negotiation may be necessary. A change process is usually included in the contract conditions.

- **Project reviews.** Regular assessment of project progress. These reviews include procurement audits where the buyer assesses the work performed by the seller and the resulting deliverables. A final review is also schedule upon project completion to ensure that all contractual deliverables have been completed and provided.
- **Contract closure.** Bringing the project to an organized close. This may include the preparation of a punch list—a list of items that has been generated by the buyer and documents any items that have not been produced satisfactorily by the seller. These items must be resolved before closure can be completed. Project closure also includes verification of contracted deliverables, formal acceptance and sign-off, a post-project review, final contract reviews, and, when appropriate, a team recognition event.

This summary of the nine knowledge areas is not meant to be a 100 percent complete listing of every item necessary to plan a project successfully. The list is designed to provide a quick reference of key items associated with the *PMBOK® Guide.*

An approach that works well when studying for the PMP exam or to just gain additional knowledge about project management is to review the overview charts of each knowledge area in the *PMBOK® Guide*. These overview charts actually provide a very thorough listing of the things project managers and project teams should be aware of. Regularly reviewing these overview charts will increase your ability to plan effectively and to determine which inputs, tools and techniques, and outputs will be needed in the particular project to which you are assigned.

Chapter Thirteen

Bringing the *PMBOK®* *Guide* to Life Through Templates

The *PMBOK® Guide* is a great resource for developing templates. The lists, descriptions, and explanations in the *PMBOK® Guide,* along with some project management knowledge and some creativity, will assist in producing customized templates for most projects and for many of the start-up deliverables associated with a well-defined and established project management office.

As an example, if there is a need to precisely define enterprise environmental factors (the higher level and unique factors internal or external to an organization that may impact how the organization is managed) for planning purposes, a table can be created. The following table shows that each item can be described in the detail required to meet the needs of the organization. These factors will vary by organization, and it is a good practice to take the time to identify and review each factor and determine how each factor may affect the project planning process.

Enterprise Environmental Factors Planning Template

Factor	*Description*
Organizational culture	Describe the high-level culture of the organization. How is the organization structured? What are the key items the team should be aware of? Describe the typical management style, openness of communication, trust, and influence of the government. Analyze the culture of each business unit and determine how the business units interact with each other. As an example consider how an engineering unit communicates and works with a sales or marketing department.
Government or industry standards	Identify the specific standards that may be associated with the project: electrical, regulatory, quality, tooling, metric, SAE, materials precision requirements, voltage, methods etc.
Infrastructure	Definition—The architectural elements, organizational support, corporate standards, methodology, data, processes, and physical hardware/network, etc. How well does the current infrastructure support project efforts? Where are the potential problems? What equipment is available? Determine the limitations of the infrastructure as well as the strengths.
Human resources	What are the skills required for the project? What is the availability of required project resources? What level of expertise exists? How well are the people performing their work? Are there morale issues to consider?

Personnel administration	What is the process for obtaining project team resources? How will team members be assessed for performance? What is the hiring and employee termination process? How is a reporting structure arranged? Who directly reports to whom?
Stakeholder risk tolerance	Describe the general view of risk management and dealing with risk within the organization. Is the organization risk averse, or does management encourage a high degree of risk taking? Determine the nature of the organization in terms of tolerance to risk issues.

PROJECT CHARTER TEMPLATE

Using the *PMBOK® Guide* description of a project charter, create an outline that will define, at a high level, the intended purpose of the project, the risks, and key planning elements of the project. The charter is usually the first step in the planning process and provides the authority to engage organizational resources. The following table shows the typical elements of a charter.

Project Charter Template

Project name
Business need
Product or service description
Project justification. What is the rationale for approving the project?
Project manager assigned. Name, location, contact information, background, and experience (as needed).
Milestones (contractual milestones or executive agreements). The use and purpose of a project charter will vary. In many cases a contract is agreed upon that generates the preparation and release of a charter. In some organizations, the charter and the contract are considered to be the same document.

Stakeholders and influences. Who are the key players and why are they involved?
Functional organizations and business units involved. This helps to prepare the organization regarding resource needs.
Assumptions. Organizational, Environmental and external. Assumptions are defined as those items that can be considered true, real, or certain for planning purposes. Examples: The organization will provide the required funding for the project. The organization has the resources capable of completing the project work. Sufficient time will be available to complete the project successfully.
Constraints. Organizational, environmental, and external. What are the known limitations? Examples: Organizational resource limitations Available funding Contractual milestones
Reference to the business case that supports the selection and justification of the project. This section can be used to describe some of the key items in the business case that resulted in the approval of the project.
Summary budget (high-level estimates). This estimate is generally an order of magnitude estimate and must be further developed during the planning process.
Approvals and Authorizations. Who will sign and approve the project and authorize the use of organizational resources?

CHECKLIST FOR MANAGING PROJECTS

Checklists are a great tool for making sure that important items have not been omitted. The *PMBOK® Guide* is actually a great checklist for a project. A review of the five process groups and overview charts for each knowledge area can provide a well-developed project planning checklist.

Create a project planning checklist by referencing the components of the *PMBOK® Guide*. Customize the checklist based on your organization's project management maturity. The details of the checklists will depend on the type and complexity of the project. The idea behind a checklist is to help to minimize the omission of critical items. A basic rule to follow is that a checklist should never be considered to be 100 percent complete. There is always something that can be added or improved.

Sample Project Planning Checklist

- ❑ A clear, concise project charter defining the project has been prepared.
- ❑ A stakeholder analysis has been completed (defines who the stakeholders are including key stakeholders and their level of influence).
- ❑ Performance objectives using SMART criteria have been written (Specific, Measurable, Attainable, Realistic, Time based).
- ❑ A project scope statement clearly defining the boundaries and constraints has been developed.
- ❑ A project kick-off meeting has been scheduled and completed.
- ❑ A work breakdown structure (WBS) has been developed to the required level of detail.
- ❑ Deliverables have been identified by phase and at the project level.
- ❑ Roles and responsibilities have been defined using a RAM (Responsibility Assignment Matrix or RACI chart (Responsible, Accountable, Consult, Inform).

- ❑ A network diagram has been developed to ensure logical dependencies have been defined.
- ❑ A project schedule has been prepared.
- ❑ A project budget has been prepared using reliable estimating techniques and data.
- ❑ A risk management plan or process has been established for the project.
- ❑ Project activity durations do not exceed the 80 hour (or 40 hour) rule.
- ❑ Mutually agreed upon monitoring and control procedures have been established.
- ❑ A plan for integrating all project components has been established.
- ❑ A change control process has been developed and communicated.
- ❑ The necessary subsidiary plans have been developed – change control, communications, quality management, safety, cut-over, etc.
- ❑ Expectations have been defined and documented for each stakeholder.
- ❑ Acceptance criteria for the project deliverables has been clearly defined.
- ❑ A recognition plan has been established for the team. (celebration of successes is a key factor in sustaining team performance).

PROJECT REVIEW OR HEALTH CHECK

Regardless of project type or complexity, it is important to conduct project reviews on a regular basis. A basic rule is to conduct a review at the end of each project phase. A template for a project review can be developed by defining the project life cycle clearly, specifying the deliverables associated with each phase, identifying any contractual agreements, and comparing project performance with existing organizational standards.

Project Name: Project Review or Health Check
SCORING FACTORS

–4 = Strongly Disagree	*–2 = Disagree*	*0 = Neutral*	*2 = Agree*	*4 = Strongly Agree*
No.	*Question*			*Score:*
1.	A well organized and sound business case supporting the need for the project has been developed and approved.			
2.	The project is clearly aligned with organizational strategy.			
3.	Senior management support has been communicated and is clearly visible for the project.			
4.	The objectives of the project are well defined using SMART criteria and have been documented and communicated to all project stakeholders.			
5.	The main project deliverable has been defined and a clear set of deliverables have been identified for each phase of the project.			
6.	The customer has participated in the project planning process, reviewed the project plans, understands his or her assigned tasks and has accepted responsibility for the assigned tasks assigned to the.			
7.	The customer has formally agreed to the scope of the project and has signed off on the plan. (Specific Project boundaries have been established to minimize scope creep and substantial change.)			
8.	Critical success factors, expectations, and performance measurement processes have been identified and agreed with the key stakeholders and the customer.			

–4 = Strongly Disagree	*–2 = Disagree*	*0 = Neutral*	*2 = Agree*	*4 = Strongly Agree*
No.	*Question*			*Score:*
9.	An escalation process has been established and communicated to all key stakeholders to ensure proper identification and resolution of project issues.			
10.	A detailed project plan, including required subsidiary plans, has been developed with input from all key stakeholders.			
11.	Contractual milestones and critical terms and conditions, including penalty clauses have been described in the project plan.			
12.	Resource estimates have been carefully determined to ensure that there are sufficient resources available for each project phase through project completion.			
13.	Roles and responsibilities have been clearly assigned using a responsibility assignment matrix (RAM). Changes and updates in responsibility have been recorded on a regular basis.			
14.	All materials required for the project have been identified and ordered to ensure availability when needed. A risk assessment has been completed to minimize potential risk events.			
15.	A risk management plan and risk register have been developed jointly with all key stakeholders to track project risks effectively.			
16.	A project communications plan has been developed and clearly defines the specific communications requirements of the project stakeholders including technology requirements and expectations regarding project status reports, format, and frequency of delivery.			

17.	A realistic project schedule has been developed using proven and effective planning techniques (reliable estimates, paralleling tasks, fast tracking, crashing use of experts).	
18.	Project success factors have been identified and communicated to all key stakeholders (success criteria has been defined beyond on time, within budget, and according to specifications.	
19.	Project status meetings are conducted on a regular basis using a standard process for meeting control – agenda, time limitations, meeting minutes distribution.	
20.	A quality assurance process is in place to verify correctness and completeness of each project deliverable.	
21.	A project team recognition plan has been established to ensure continued high performance.	
	Score	

Project Health Check – Scoring Criteria

Score	*Probability of Success*	*Action Required*
–84 to –42	Project success is e unlikely; there are significant issues and risk situations that require immediate attention.	Identify and Analyze all issues and risk areas. Escalate issues that require executive attention. Obtain suggestions from peers and others who may have experience with the identified issues. Prioritize the major issues and identify root causes. Prepare a report for project executives, explain the issues and a plan for recovery.

Score	*Probability of Success*	*Action Required*
>–42 to –0	Low	There is significant uncertainty about this project. Important information may not be available or has not been discussed. Assess each low scored item and develop a specific response. Identify areas of strength and confidence to communicate these items to the project team. Develop plans to address weak areas and threats.
>0 to 42	Moderate	Project performance can be improved. Identify and assess risks associated with each item that received a low score determine the appropriate response, obtain input and support from the project team or functional units, and communicate the condition of the project to the sponsor and project team.
>42 to –84	High	The project manager has established a strong cohesive team, a well organized plan is in place, project status is reported in a timely manner and with efficiency. Continue to monitor the project, schedule recognition events for the team to maintain morale, discuss opportunities for continuous improvement, ensure that the project sponsor is aware of the project condition.

PROJECT SCOPE STATEMENT

A project scope statement is a critical element of the project planning process. The *PMBOK® Guide* provides a detailed list of the key elements of a project scope statement. A template can be developed for use by the project team, project manager, or the specific person or group designated and assigned to write the scope statement, The scope statement, as a general rule, should be a document that clearly defines the project in sufficient detail, and provides the basis for more complete planning of the project. The scope statement is not a plan, but it is essential in the planning process.

Project Scope Statement Example

Project Name: ______________________
Project Start Date: ______________________
Project Manager: ______________________
Customer / Client: ______________________
Project Sponsor: ______________________

Executive Summary is a brief description of the project. This information may be obtained from the project charter.

Executive Summary

Project Objectives: Measurable success criteria. (What must be accomplished? When? What is the cost?) Examples: Deliver system, train end users, distribute software, deploy new methodology, build new test lab. Use endpoint words: train, distribute, increase, reduce, organize.	
Product scope definition: Specific characteristics of the product that the project was initiated to produce. (What will be produced? Features, functionality, physical characteristics. Provide information about the amount of work that will be required.	

Project requirements. Who will be involved in the project? Who will receive the deliverables or final product? (Needs of the stakeholders, conditions that must be met, capabilities required to meet a contractual specification.) In this context, the project requirements are those items needed by the team to deliver the final product. Requirements may include—ability to report project resource time, ability to track project budget, training programs for the team, appropriate tools and equipment, and support systems.	
Project boundaries. What is included in the project and what is excluded from the project? These are not constraints or limitations. Project boundaries are those items that will be produced as a result of the project. Items not included in the scope of the project may be described in the project documentation to clarify the project scope of work and minimize the possibility of miscommunication and misinterpretation.	
Project deliverables. The product or products that will be produced and other ancillary items such as reports and documentation. Define or explain the final product or products of the project. (These are the tangible items that will be produced.) Examples: a new system with all peripherals, a book, and training course with all supporting materials.	

Product acceptance criteria. (What specifically must be provided for the intended customer to accept the product?) The process for accepting project deliverables. This may include the desired quality of the product, appearance, packaging, functionality, ease of use, availability, maintainability, and others.	
Project constraints. Examples of constraints include Budget limitations, schedule milestones, imposed dates, and agreed-upon contractual.	
Project assumptions. Assumptions are those items that can be considered to be real, true, or certain for planning purposes. An assumption set may be provided during the project start up process. These assumptions should be reviewed and validated. Determine the potential risks and impact to the project if assumptions are found to be false or incorrect. Examples of assumptions: Sufficient resources will be available. The project budget will be fully supported by management. There will be sufficient management support for the project.	
Project organization. How will the project staff and team be organized? Include Project team and other stakeholders.	

Initial defined risks. Identify potential risk events that may impact planning and execution. Consider technical risks, environmental risks, staffing issues, resource capability, other projects in progress.	
Project funding limitations. This is the total allowable amount authorized for the project. Identify differences between the amount of funding made available for the project and the actual amount that may be required.	
Cost Estimate. This is the Initial cost estimate for the project that is prepared prior to detailed planning.	
Project configuration management requirements. These are the change control processes that will be implemented through the project life cycle to manage product characteristics (size, features, functions).	
Project Specifications – The compliance requirements associated with the product to be delivered. Examples: certifications, contractual testing specifications, safety testing.	
Approval requirements. Processes for approval of project objectives, deliverables, documents, work results. The conditions that must be met to obtain approval, including testing, reviews and who is authorized to approve project related work.	

PROJECT JEOPARDY REPORT—REPORTING THE "TROUBLED PROJECT"

Occasionally, projects become *troubled*. This means that things did not go according to plan, there is a slippage in schedule, an overrun of the budget, or some unplanned situation has developed that requires immediate attention by executive management or sponsors. Part of the monitoring and control process that is usually communicated at the start of the project includes developing some type of escalation plan or process for involving management when problems arise. A project **jeopardy report** can be developed to facilitate the process. This form is especially useful in the virtual project environment. Again, the *PMBOK® Guide* provides the foundation for developing a process to identify and resolve project problem situations.

Managing Projects in Trouble

The project manager at team should have a plan in place to manage projects that deviate from approved plans. A project jeopardy reporting process is useful in managing the communication and escalation of project issues.

Jeopardy is defined as follows, based on how the situation affects the project. Generally, there are different levels of severity:

Level 1 jeopardy. The project due date is in immediate danger. A significant problem exists.

Level 2 jeopardy. The project due date may be impacted. This type of jeopardy frequently is elevated to level.1

Commitment jeopardy. A noncritical task may not be completed as planned. The project due date is not affected (at this time).

Sample Jeopardy Report

Project name: ____

Project manager: ______________________________________

Project sponsor: ______________________________________

Brief project description: ______________________________

Report Submitted by:

____

Description of jeopardy item:

____

____

Action required to relieve or resolve the jeopardy:

____

____

Ramification if the jeopardy situation is not relieved:

____

____

Required response date: _________________________________

Person responsible and organization: _____________________

Responsible 3rd level mgr.: ______________________

Telephone: ______________________

____

Responsible 2nd level mgr.: ______________________

Telephone: ______________________

____

Project manager: ______________________

Telephone: ______________________

____

When you distribute the report, include only those stakeholders who have a need to know about the project jeopardy situation or will be involved in resolving the issue. Minimize the distribution to avoid involvement by unneeded parties.

CLOSURE WORK BREAKDOWN STRUCTURE

Closing a project can be a project in itself. A WBS can be developed and added to the project plan to ensure a smooth and well-managed transition from the project environment to the operations environment. Review the inputs, tools and techniques, and outputs of the contract closure process in the *PMBOK® Guide* and create a WBS that will guide the project team through to a successful project closeout.

Example Project Closure WBS

1.0 PROJECT CLOSURE

1.1 CONTRACTUAL

1.1.1 Internal

Verify and validate completed project deliverables.
Identify remaining deliverables.
Close open change requests.
Finalize project financials.
Identify certification requirements.
Close open work orders.
Communicate closures.
Obtain client punch list.
Resolve client punch list.
Prepare as-built drawings.
Finalize as-built drawings.
Resolve approved work-arounds.
Clean up physical facilities.
Hand back to client.
Conduct post implementation review.

1.1.2 External

Confirm remaining deliverables with client.
Obtain closure of open contract changes and billing.

Obtain closure of open contractor work orders.
Obtain closure of open change orders.

1.2 ORGANIZATIONAL

1.2.1 Technical

Critique system or hardware performance.
Critique system or hardware technical effort.

1.3 ADMINISTRATIVE

Identify physical facilities for disconnect or disposal.
Dispose of or redeploy physical facilities/equipment.
Complete and distribute project financials and project documentation.
Provide project documentation to customer, sales, or services organization.
Provide training materials to client or receiving organization.

PROJECT QUALITY PLAN OUTLINE

All projects should include a quality plan to ensure consistent levels of performance in each phase and to ensure acceptable completion of project deliverables. Quality plans will vary, depending of the culture of the organization, the type of product involved, and the general quality policy of an organization. Before developing a quality plan for the project, make sure the team is aware of and understands the quality policy that has been developed and communicated by the leaders of the organization.

Quality Policy Example

We, the employees of ______ (company or organization name), are committed to understand, meet and, when possible, exceed our Customer's Requirements and expectations through the continuous improvement of our processes. We are dedicated to delivering defect-free products and services on-time at a competitive cost, and with customer satisfaction as our driving principle.

A project quality policy, based on the organization's overarching policy could be developed by the project team during the project start up or kickoff meeting. This establishes the importance of delivering high quality products and creates a greater sense of overall pride and commitment to the project.

QUALITY PLAN OUTLINE

Customer—Provide the name of the main customer (Identify the individual or organization receiving the product.)
Stakeholders—identify the key stakeholders? (Key stakeholders may also be considered customers if they will be receiving some type of deliverable during the project life cycle. The key stakeholders should be clearly documented and agreed upon.

Examples of Stakeholders:

- Product manager
- Technical planners
- SITE Managers
- Business unit – representatives
- End-users (users of the products and services)
- Project team members
- Project manager
- Sponsor
- Suppliers / vendors
- Customers

Document the expectations of the main customer (In terms of deliverables, performance, communications, service levels, etc.)

Examples of Customer Expectations:

- Transparency of service (no noticeable disruptions in operations)
- Minimal down time of systems
- Appropriate Training provided to users

- No increase in "normal" volume of trouble reports
- Project completed on schedule
- Project manager and team maintains sensitivity to changing needs of the customer
- Back up or Contingency plans in place
- Regularly scheduled and effective status reports

How will the customer's expectations be identified and addressed?

Methods for defining Expectations:

- Conduct a needs analysis to identify concerns and critical performance areas
- Define and manage interface agreements between business entities
- Provide training in the use of quality processes
- Establish acceptance criteria for all deliverables
- End of phase reviews
- Walkthroughs and inspections

How will the project team measure quality and project success?

Examples of Measures:

- Service levels
- Error rate / defect rate
- Product sampling
- Security levels
- System response time
- Percent of defects
- Timeliness of corrective actions

Identify the key elements that could affect the quality of the project

- Organizational—Environmental factors and process assets
- National—Quality norms at the national level (industry standards, benchmarks

- Regional—Quality norms at the regional or local level
- International—Customs, interpretation of quality, political issues
- Project quality goals—Specific project quality goals that will define the success of the project:
  - 100 percent customer satisfaction
  - Continuous process improvement through lessons learned
  - Reduction in failure rate
  - Reduction in scrap and rework
  - Reduction in setup time
  - Manufacturing cycle time

RISK MANAGEMENT

Risk management is a critical factor in the pursuit of project success. There are many ways to identify potential project risks, and the *PMBOK® Guide* can be a great source for developing templates that will assist in the identification and management of project risks.

Project Risk Checklist

Category	*Description*	*Investigate*	
		Yes	*No*
Requirements	Are there any TBDs in the specifications? Are the external interfaces completely defined? Do you and the client have the same understanding about features and functionality? Is a requirements management process in place? Is a requirements change management plan in place?		

Category	*Description*	*Investigate*	
Design	Are there phase overlaps or is fast tracking applied? Are design reviews scheduled in the project plan? What are the potential design problems? What could go wrong with the design? What incompatibilities may exist?		
Resources	Are the required resources available? • Quantity • Competency levels		
Integration and Testing	What integration issues may create problems? How well is the integration testing process defined? What are the specific integration test requirements? Does the team have the capability to perform the required tests?		
Management Methods	Is the appropriate training available for the project team and end users? What is the quality assurance process? Are control procedures in place for changes?		
Quality	Is a quality plan in place?		
Subcontractors and suppliers	How has each contractor been assessed for capability and financial stability?		
Technology	What technology problems may be expected? • Interface with existing systems • Legacy systems • Inter-operability • Ease of use • Reliability		

Project Risk Register Template

Deliverable, Phase or Component	*Potential Risk Events*	*Impact*	*Prob.*	*Risk Rating*	*Reasons For Risks*	*Urgency*	*Response or Counter-Measures*	*Person Assigned*	*Status / Final Resolution and Date*

Establishing categories for risks facilitates the risk identification process and helps to create an ongoing and continuously developing list of risks that can help the project team respond to risk effectively and confidently.

Source of Risk Template

Risk Category	*Sources of Risk*
Technical—Evolving design, reliability, operability, maintainability	Physical properties of the product or system Changing requirements Material properties Unstable technology Unreliable testing methods for new technology Modeling system complexity Integration testing Design flaws Safety issues due to untested technology
Program level and Organizational—Processes for obtaining resources. Enterprise environmental factors, Organizational process assets	Material availability Contractor stability Personnel availability Regulatory changes Personal skills Organizational process Security processes Communications processes
Supportability— ability to maintain systems Operating procedures	Product reliability Training effectiveness System safety Documentation—quality and availability Technical data—quality and availability Interoperability Transportability

Cost—Limited budgets, effectiveness of estimating processes, project constraints	Administrative rates Unplanned Overhead costs Estimating errors Cost of quality—cost of repairs and rework Reliability of estimating resources
Schedule—Estimating processes, reliability of planning processes, methods and procedures	Estimating errors Number of critical path items Degree of Concurrency Unrealistic schedule baseline

The *PMBOK® Guide* is a great source for obtaining the information needed to develop project templates. The amount of detail included in a template depends on the project and the needs of the project stakeholders. Templates are useful tools and can significantly improve the quality of the planning process and can accelerate project planning. Templates create excellent lessons-learned files and can always be improved. Virtually any item in the *PMBOK® Guide* can be used to create a template. Just select the planning item needed for the project, make sure you understand why the item is needed and how it will be used as the plan is developed

Example: Monitoring and Control Template: Work Performance Measures

1. Schedule progress.
2. Deliverables produced to date.
3. Activities in progress.
4. Activities scheduled but not started.
5. Activities completed.
6. Level of quality achieved.
7. Costs planned and costs incurred.
8. Estimate to complete.

9. Percent complete—activity level, task level, summary task level, project level.
10. Resource utilization.
11. Project performance forecasts—Estimate at complete (EAC).

Determine what you wish to accomplish, create an outline that provides a description of what is needed, determine the actions required to achieve the desired result, and you have a template.

Creating templates will broaden your knowledge of project management, facilitate the development of a well-organized and complete project plan, minimize confusion among project stakeholders, and create an excellent repository of information for the entire enterprise. Templates accelerate the planning process and will definitely help to bring the *PMBOK® Guide* to life!

References

CHAPTER 1

Project Management Institute. *A Guide to the Project Management Body of Knowledge*. Newtown Square, PA: Project Management Institute, 1996.

Project Management Institute. *A Guide to the Project Management Body of Knowledge,* 2nd ed. Newtown Square, PA: Project Management Institute, 2000.

Project Management Institute. *A Guide to the Project Management Body of Knowledge*, 3rd ed. Newtown Square, PA: Project Management Institute, 2004.

CHAPTER 2

State of Texas Department of Information Resources. "Project Charter." www.dir.state.tx.us/pubs/framework/gate1/projectcharter, accessed November 2008.

CHAPTER 4

Johnson, Jim, Karen D. Boucher, Kyle Connors, and James Robinson. "Collaborating on Project Success." *Softwaremag.com* (February/ March 2001). www.softwaremag.com/archive/2001feb/collaborativemgt.html, accessed October 2008.

Project Management Institute. *A Guide to the Project Management Body of Knowledge*, 3rd ed. Newtown Square, PA: Project Management Institute, 2004.

Wysocki, Robert, James P. Lewis, and Doug DeCarlo. *The World Class Project Manager,* New York: Perseus Publishing, 2001.

CHAPTER 5

Wysocki, Robert, James P. Lewis, and Doug DeCarlo. *The World Class Project Manager*. New York: Perseus Publishing, 2001.

Developing Performance Measures—A Systematic Approach.

CHAPTER 8

Kerzner, Harold. *A Systems Approach to Planning, Scheduling, and Controlling*. 9th ed. Hoboken, NJ: John Wiley & Sons, 2006.

CHAPTER 9

Kerzner, Harold. *A Systems Approach to Planning, Scheduling, and Controlling*. 9th ed. Hoboken, NJ: John Wiley & Sons, 2006.

Project Management Institute. *A Guide to the Project Management Body of Knowledge*, 3rd ed. Newtown Square, PA: Project Management Institute, 2004.

Martin, Paula, and Karen Tate. "Team-Based Risk Assessment—Turning Naysayers and Saboteurs into Supporters." *PM Network* magazine (February 1998).

CHAPTER 10

DRM Associates and PD-Trak Solutions. "New Product Development Body of Knowledge." www.npd-solutions.com/bok.html, accessed March 2008.

Project Management Institute. A Guide to the Project Management Body of Knowledge®, 4th ed. Project Management Institute, 2008.

Martin, Paula, and Karen Tate. "Team-Based Risk Assessment—Turning Naysayers and Saboteurs into Supporters." *PM Network* magazine (February 1998).

Index